Alternative Sources of Energy Modeling and Automation

Alternative Sources of Energy Modeling and Automation

Special Issue Editor
George S. Stavrakakis

MDPI • Basel • Beijing • Wuhan • Barcelona • Belgrade • Manchester • Tokyo • Cluj • Tianjin

Special Issue Editor
George S. Stavrakakis
Technical University of Crete
Greece

Editorial Office
MDPI
St. Alban-Anlage 66
4052 Basel, Switzerland

This is a reprint of articles from the Special Issue published online in the open access journal *Energies* (ISSN 1996-1073) (available at: https://www.mdpi.com/journal/energies/special_issues/alternative_sources_energy_modeling_automation).

For citation purposes, cite each article independently as indicated on the article page online and as indicated below:

LastName, A.A.; LastName, B.B.; LastName, C.C. Article Title. *Journal Name* **Year**, *Article Number*, Page Range.

ISBN 978-3-03928-374-3 (Pbk)
ISBN 978-3-03928-375-0 (PDF)

© 2020 by the authors. Articles in this book are Open Access and distributed under the Creative Commons Attribution (CC BY) license, which allows users to download, copy and build upon published articles, as long as the author and publisher are properly credited, which ensures maximum dissemination and a wider impact of our publications.

The book as a whole is distributed by MDPI under the terms and conditions of the Creative Commons license CC BY-NC-ND.

Contents

About the Special Issue Editor vii

Preface to "Alternative Sources of Energy Modeling and Automation" ix

Józef Rak, Przemysław Błasiak and Piotr Kolasiński
Influence of the Applied Working Fluid and the Arrangement of the Steering Edges on Multi-Vane Expander Performance in Micro ORC System
Reprinted from: *Energies* **2018**, *11*, 892, doi:10.3390/en11040892 1

Alexandros Arsalis and George E. Georghiou
A Decentralized, Hybrid Photovoltaic-Solid Oxide Fuel Cell System for Application to a Commercial Building
Reprinted from: *Energies* **2018**, *11*, 3512, doi:10.3390/en11123512 17

Kevin Ellingwood, Seyed Mostafa Safdarnejad, Khalid Rashid and Kody Powell
Leveraging Energy Storage in a Solar-Tower and Combined Cycle Hybrid Power Plant
Reprinted from: *Energies* **2019**, *12*, 40, doi:10.3390/en12010040 37

Dimitris Al. Katsaprakakis and Georgios Zidianakis
Optimized Dimensioning and Operation Automation for a Solar-Combi System for Indoor Space Heating. A Case Study for a School Building in Crete
Reprinted from: *Energies* **2019**, *12*, 177, doi:10.3390/en12010177 61

Efstratios Batzelis
Non-Iterative Methods for the Extraction of the Single-Diode Model Parameters of Photovoltaic Modules: A Review and Comparative Assessment
Reprinted from: *Energies* **2019**, *12*, 358, doi:10.3390/en12030358 83

Polychronis Spanoudakis, Nikolaos C. Tsourveloudis, Lefteris Doitsidis and Emmanuel S. Karapidakis
Experimental Research of Transmissions on Electric Vehicles' Energy Consumption
Reprinted from: *Energies* **2019**, *12*, 388, doi:10.3390/en12030388 109

Manuel Angel Gadeo-Martos, Antonio Jesús Yuste-Delgado, Florencia Almonacid Cruz, Jose-Angel Fernandez-Prieto and Joaquin Canada-Bago
Modeling a High Concentrator Photovoltaic Module Using Fuzzy Rule-Based Systems
Reprinted from: *Energies* **2019**, *12*, 567, doi:10.3390/en12030567 125

Neofytos Neofytou, Konstantinos Blazakis, Yiannis Katsigiannis and Georgios Stavrakakis
Modeling Vehicles to Grid as a Source of Distributed Frequency Regulation in Isolated Grids with Significant RES Penetration
Reprinted from: *Energies* **2019**, *12*, 720, doi:10.3390/en12040720 147

About the Special Issue Editor

George S. Stavrakakis (Prof.) received his first degree in Electrical Engineering from the N.T.U.A. (National Technical University of Athens), Athens, Greece, in 1980. His postgraduate D.E.A. in Automatic Control and Systems Engineering was obtained from I.N.S.A., Toulouse, France in July 1981 and his Ph.D. in the same area was obtained from Paul Sabatier University-Toulouse-III, Toulouse, France in January 1984. He has worked as a research fellow in the Robotics Laboratory of N.T.U.A., Athens (1985–1988) and as a visiting scientist at the Institute for System Engineering and Informatics/Components Diagnostics & Reliability Sector of the EC—Joint Research Centre (JRC) in Ispra, Italy (September 1989–September 1990). He was vice president of the Hellenic Center of Renewable Energy Sources (CRES) November 2000–April 2002. In September 1990, he joined the Electronic and Computer Engineering Department, Technical University of Crete, Chania, Crete, Greece as an associate professor. He is currently a full professor (from November 1995) in the same department, Electric Circuits and Renewable Energy Systems Laboratory (www.elci.tuc.gr). He performs prototype and advanced engineering research with applications in: robotics, process re-engineering, systems safety and reliability analysis, real-time industrial processes fault monitoring and diagnosis, modeling and diagnosis in bioengineering systems, informatics and electronics and microcomputer applications in power systems, renewable energy sources (RES) modeling and automation and autonomous power systems, smart and micro grids, energy storage and autonomous RES integration and optimal operation, increased RES penetration in non-interconnected power grids, energy efficiency and building energy management systems (BEMS), solid waste management, and waste-to-energy technologies. He is the author or co-author of about 150 full papers published in international refereed journals, international edition books, and refereed conferences, about 65 research reports, 5 international edition and dissemination scientific books, and 70 papers in international non-refereed conferences on the above topics. He has participated in 39 European Commission and national funded research projects, being the coordinator of 9 among them. He is the author of the chapter "Electrical Parts of Wind Turbines" in the award winning handbook "Comprehensive Renewable Energy" published by Elsevier in May 2012. He is a permanently invited reviewer in the following high impact factor international scientific journals: IEEE Transactions on Energy Conversion, Electric Power Systems Research, International Journal of Distributed Sensor Networks, Renewable Energy, IEEE Transactions on Automation Science and Engineering, Energies (member of the Editorial Board), Applied Sciences, and Journal of Intelligent and Robotic Systems. He speaks and writes excellent Greek (native), with very good English, French, and Italian. He is married and has two daughters.

Preface to "Alternative Sources of Energy Modeling and Automation"

This Special Issue entitled Alternative Sources of Energy Modeling and Automation is proposed for the international journal Energies to cover original research and scientific contributions related to the formulation of computer implemented models of alternative (renewable) sources of energy (ASE/RES), which are essential for the proper allocation of widely available renewable energy sources. These models are necessary to design and implement efficient automation for the optimal operation of ASE/RES plants and installations. Detailed simulations of alternative-source energy devices and integrated power plants are cost effective solutions. Often, several subsystems of an integrated ASE/RES power plant might be inappropriate, difficult to find, and/or very expensive. ASE/RES forecasts are essential to the integration of renewable power generation in electricity markets operations, since markets ought to be cleared in advance, then market participants can make appropriate decisions. The above-mentioned topics also include the most common ASE/RES small electric power plants, such as wind, photovoltaic, solar thermal, passive solar, energy savings, small hydro, geothermal, biomass, tidal, fuel cells, batteries, hybrid plants, and electric vehicles. As an example, to design integrated smart power grids based on ASE/RES with battery-based storage, a high quality behavioral model of the ASE/RES components and automation of the integrated system are required. All the contributions of the Special Issue present innovative theoretical and applied results, methods, and solutions on the above topics.

George S. Stavrakakis
Special Issue Editor

Article

Influence of the Applied Working Fluid and the Arrangement of the Steering Edges on Multi-Vane Expander Performance in Micro ORC System

Józef Rak, Przemysław Błasiak and Piotr Kolasiński *

Department of Thermodynamics, Theory of Machines and Thermal Systems,
Faculty of Mechanical and Power Engineering, Wrocław University of Science and Technology, Wybrzeże Wyspiańskiego 27, Wrocław 50-370, Poland; jozef.rak@pwr.edu.pl (J.R.); przemyslaw.blasiak@pwr.edu.pl (P.B.)
* Correspondence: piotr.kolasinski@pwr.edu.pl; Tel.: +48-71-320-23-39

Received: 22 March 2018; Accepted: 10 April 2018; Published: 11 April 2018

Abstract: Micro-power domestic organic Rankine cycle (ORC) systems are nowadays of great interest. These systems are considered for combined heat and power (CHP) generation in domestic and distributed applications. The main issues of ORC systems design is selection of the expander and the working fluid. Thanks to their positive features, multi-vane expanders are especially promising for application in micro-power ORC systems. These expanders are very simple in design, small in dimensions, inexpensive and feature low gas flow capacity and expansion ratio. The application of multi-vane expanders in ORC systems is innovative and currently limited to prototype applications. However, a literature review indicates the growing interest in these machines and the potential for practical implementation. For this reason, it is necessary to conduct detailed studies on the multi-vane expanders operation in ORC systems. In this paper the results of experimental and numerical investigations on the influence of the applied working fluid and the arrangement of the steering edges on multi-vane expander performance in micro ORC system are reported. The experiments were performed using the specially designed lab test-stand, i.e. the domestic ORC system. Numerical simulations were proceeded in ANSYS CFX software (ANSYS, Inc., Canonsburg, PA, USA) and were focused on determining the expander performance under various flow conditions of different working fluids. Detailed numerical analysis of the arrangement of the machine steering edges showed existence of optimal mutual position of the inlet and outlet port for which the multi-vane expander achieves maximum internal work and internal efficiency.

Keywords: ORC; working fluid; multi-vane expander; numerical analysis; experimental analysis

---

1. Introduction

The operation principle of organic Rankine cycle systems (ORCs) is the same as classic Clausius-Rankine (CR) steam power plants. However, the design of these systems is different. ORCs are usually featuring smaller overall dimensions and system power. In ORC systems, heat exchangers of special design (i.e., hermetically sealed evaporator and condenser) are applied instead of steam boiler and condenser. Classically applied steam turbine does not meet the requirements of low-boiling gas, thus it has to be replaced with a specially designed turbine or volumetric expander. ORCs are mainly powered by the heat harvested from renewable (e.g., solar, geothermal) and waste sources (e.g., industrial waste heat). Using ORC system, this energy can be then converted into electricity, heating, and cooling. Different temperature level of the heat carrier can be applied. Due to their large share, low (40–250 °C) and medium temperature (250–500 °C) carriers are especially promising for powering ORC systems. Different features can be taken into account to classify ORC systems. When the power of system is considered, ORCs can be classified into high power (500 kW or more), medium power (100–500 kW), small power (10–100 kW), and micro power (0.5–10 kW) systems. ORCs can be powered by high-, medium-, and low temperature heat sources [1,2]. Turbines and volumetric expanders can be adopted as the expansion machine. The modern energy market is characterized by the diversification of the energy supply. Reliable and safe small- and micro-power ORC systems (e.g., for domestic applications) are fitting well to the concept of dispersed energy generation and thus are currently intensively developed in a number of R&D and scientific institutions around the world. Despite intensive studies these systems are still not fully developed and there are no commercially viable solutions currently available. The majority of these systems are at the level of lab-prototypes. The operating conditions of domestic, small- and micro-power ORC systems depend on the heat source's nature, i.e., its thermal power (which in the case of domestic systems is often low), output and temperature characteristics (which are often floating). Moreover, the heat carrier is often featuring low temperature. These changing conditions of the heat supply may negatively influence the ORC system operation. By the technical configuration domestic ORC systems can be classified into power plants and CHPs. In domestic ORCs micro turbines or volumetric machines can be applied as the expander. The main research topics on domestic ORCs are the selection of working fluid and the optimization of design of heat exchangers, pumps and expanders. Domestic ORCs should be simple design, cheap, safe and reliable. The external dimensions of such systems should also be minimized. The ORC system cost mostly depend on the price of heat exchangers and the expander. When the features of expander are analysed it can be pointed out that micro turbine requires a high and stable level of the working fluid flow in order to provide high efficiency and optimum operating conditions. Such flow conditions are obtained when high-output and high-power pumps are applied, what stays in contrary with the desired simplicity of the domestic ORCs. Moreover, small external dimensions of the micro turbine result in very high rotational speeds of the rotor leading to complications in the connection of the expander and electrical generator. The high precision workmanship is required in micro turbine design what directly translates into the high price of such machines. Currently, micro turbines dedicated to domestic ORC systems are still under research and development.

Research programs on ORC-dedicated micro turbines were conducted, i.a., at the Institute of Fluid Flow Machinery, Polish Academy of Sciences, Gdansk, Poland. Subsequently, such research was proceeded at the Mechanical Faculty of Gdansk University of Technology. These comprehensive studies included theoretical, experimental, and numerical modelling of fluid flow, thermodynamic phenomena, and mechanical strength in ORC-dedicated micro turbines. The prototypes of axial-flow and radial-flow micro turbines were designed, developed and successfully tested. The results of these surveys were presented in a number of books [3–8] and papers [9–18]. However, despite the successful experimental investigations, currently there are no commercially viable ORC-dedicated micro turbines yet available.

Compared to micro turbines, volumetric expanders feature a lower range of operating pressure and lower gas flow capacity. One of the criteria of selecting the volumetric expander to ORC system is the expansion ratio, i.e., the ratio of the working fluid pressure at the inlet and at the outlet of the expander. The different types of volumetric expanders are featuring different achievable expansion ratios. It is worth noting that the value of the highest and the lowest working fluid pressure in the ORC system depends on different factors. Apart from the heat source and the heat sink temperature the type of selected working fluid also influences the level of pressures achievable inside the system. Thus, different types of expanders may achieve different level of power while different working fluids are considered. In the prototypes of ORC systems (featuring small and micro-power) mostly volumetric expanders are adopted (e.g., scroll and piston). Each of the mentioned expander has different features. Scroll expanders are often used. However, they are complex in design (the maintaining of high quality of scroll workmanship is needed). Thus, their manufacturing requires advanced engineering facilities, which directly translates into a high price for such machines. The benefits are connected with the lack of valves and smooth operation resulting from existence of many working chambers in one period of time. Prototypes of scroll expanders adopted in ORC systems are, in most of the cases, reversed scroll compressors manufactured as oil-free versions. The achievable expansion (pressure) ratio in case of scroll expander is lower than 11 [19]. The field of application of scroll expander in micro ORC systems is still being investigated. The issues related to the application of scroll expanders in micro ORC systems were comprehensively treated in [19–26]. The piston expander has a much simpler design than a spiral one, but requires lubrication, valve timing; and as a result of its cyclical operation, pressure pulsations, noise, and vibrations are generated. The experimental results related to piston expanders were presented in [19,27,28]. On the other hand, some research results [29] indicate that rotary lobe expanders are promising for application in small-scale power systems. However, these expanders are still in the very early stage of development.

In turn, the multi-vane expander design is very simple, which translates into low production costs. The multi-vane volumetric machines are currently manufactured in a number of factories worldwide, and they are successfully applied in different branches of the industry as e.g., pneumatic motors, compressors, and vacuum or liquid pumps. The design of the multi-vane machine together with the principle of its operation was comprehensively described in [30].

The design of multi-vane expander shows many positive features indicating that it can be promising alternative to other volumetric expanders and micro turbines applied in ORCs. This includes an advantageous ratio of the power output to the external dimensions, a lower gas flow capacity and lower expansion ratios compared to the other types of volumetric machines and micro turbines. When special construction materials are applied the need for lubrication can be eliminated. The multi-vane expander can be also easily hermetically sealed, which is one of the key issues when the safety of the system is considered. Multi-vane machines can expand the gas–liquid mixture without serious problems. The achievable power output of multi-vane expanders (several hundred W to approximately 5 kW) fits well to the desired power ranges of domestic power systems. The low pressure ranges (maximum gas pressure on the inlet to machine of ca. 10 bar) and low rotational speeds (of ca. 3000 rev/min) are also advantageous.

The above-mentioned positive features of the multi-vane expanders enable application of these machines in small- and micro- power domestic ORC systems powered by different heat sources, such as biomass, solar, waste, or geothermal heat.

The literature review shows that research in the field of application of multi-vane machines to micro ORC systems is growing and it is devoted to different topics. The first one includes the design issues and experimental works [2,30–34]. Second is the numerical analysis of the expanders operation and optimization of their design [35–40]. Some works are dedicated to the design of multi-vane pumps that can be applied in ORC systems [41]. Issues that are still insufficiently recognized and scientifically described are related to different topics. Firstly, the operating of multi-vane expander with different low-boiling working fluids should be comprehensively investigated. Modelling works should be proceeded also on limiting the effect of internal leakages of working fluid in expanders' working chambers as well as on the lowering the expanders' internal friction. The optimization of the arrangement and the design of the inlet and outlet ports together with works on the optimization of the expander geometry are also necessary. At last, a modelling of gas–liquid mixture expansion should be investigated in more details. The solutions for the raised problems require comprehensive theoretical study and experimental analysis. Thus, the authors decided to carry out the experimental and numerical study on the influence of the applied working fluid and the influence of the arrangement of the steering edges on operating conditions of multi-vane expander in micro ORC system.

In the following sections, the experimental results together with the description of the multi-vane expander and test-stand are presented, followed by results of numerical analysis.

Initially, the numerical model of the expander was established. Then, the model was validated using experimental data. This way its suitability was proved. Obtained model was then applied for testing modifications to machine design and finding the optimum working parameters, working fluid, and steering edges arrangement. Different working fluids were considered during the numerical analysis and the expander operation was modelled for a range of pressure and temperature parameters.

The differences in the multi-vane expander operation for these working media were comprehensively investigated. Basing on the numerically obtained results, the optimal working point for the organic Rankine cycle was found. Moreover, the numerically obtained results gave the insight in the flow phenomena in the expanders' working chambers. The differences in these phenomena for analysed working fluids were investigated. Furthermore, the numerical analysis indicated the imperfections of the expanders' design. It was found that the sharp edges of the inlet and outlet ports are causing working medium turbulence and the internal leakages are reducing expanders' output power.

2. Experimental Results and Numerical Model

2.1. Description of the Experimental Test-Stand, the Multi-Vane Expander, and the Experiment Results

The experimental investigations were carried out using a prototype of domestic combined heat and power (CHP) ORC system utilizing the multi-vane expander and featuring the thermal power of 18 kW and electric power of 300 W.

The test-stand is adopting an air motor (featuring the maximal power of 300 W) as the expander. The air motor was adapted for low-boiling working fluid. Namely stainless steel bearings, PTFE seals, and graphite vanes were applied. Figure 1 shows a general view of the disassembled expander and its main components i.e., the cylinder (1), the rotor (2), vanes (3), two end covers (4), rolling bearings (5) and vane guiding rings (6). The cylinder has the inner diameter of 37.5 mm and the length of 22.0 mm. The rotor which is mounted eccentrically in the cylinder (eccentricity is equal to 1.75 mm) has the outer diameter of 34.0 mm. The rotor has milled perpendicular slots in which the vanes are moving. Vanes are guided with help of the rings. The rotor is supported by rolling bearing which are mounted in the end covers. The expander is fed with the gas through the inlet port. After the expansion the gas exits the machine thorough the outlet port. The inner diameter of the inlet and the outlet port is equal to 8.5 mm.

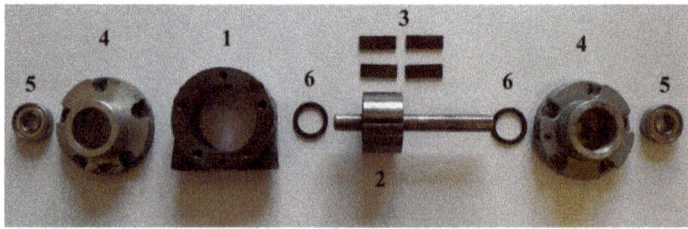

Figure 1. A general view of the expander. (1) cylinder; (2) rotor; (3) vanes; (4) end covers; (5) rolling bearings; (6) rings.

The experimental series were performed in order to obtain the data needed for validation of expanders' numerical model. Therefore, varied measurement conditions were applied on the test-stand. The values of the heat source temperature were varied between 55 °C and 85 °C and the working fluid pressure was varied between 2.0 bar and 5.2 bar. The experimental results showed that the expander internal work ($l_i = h_{in} - h_{out}$) (where h_{in} and h_{out} is the enthalpy of the working fluid at the inlet and at the outlet of the expander) varies in the range of 0.96–4.18 kJ/kg while its internal efficiency $\left(\eta_i = \frac{h_{in} - h_{out}}{h_{in} - h_{outs}}\right)$ (where h_{outs} is the enthalpy of the working fluid at the outlet of the expander when the isentropic expansion process is considered) varies in the range of 17.2–58.3% depending on the experimental conditions.

The expanders' numerical model (see Section 2.2) was built based on the measurements of the expander dimensions used during laboratory tests. The obtained numerical model was then validated using the experimental data. The results from the experiment were found to be in good agreement with the numerical predictions. Thus, it was concluded that the numerical model is validated for a given range of temperature and pressure parameters.

In the following paragraphs, the results of numerical simulations are presented. A more detailed description of the experiment and experimental results is presented in [31,37].

2.2. Description of the Numerical Model

The movement of the multi-vane expander is inherently transient and its modelling required special treatment via transient boundary conditions. The model takes into account the rotational motion of the rotor and the vanes. For each time step the vanes radial position change hence the new topology of the numerical mesh is calculated and updated based on the equations of motion of the multi-vane expander. This allows for solution of the pressure, velocity and temperature fields inside a working chamber and provides precise information on the thermo-flow phenomena in a multi-vane expander, such as the internal leakages. The inlet and the outlet pipes remain stationary and they are connected with the rotor domain through the interface (see Figure 2). In each time step, the mesh was moved according to the following algorithm:

1. An angular position of a vane defined by θ angle is calculated.
2. A vane is rotated by an angle $-\theta$ to be aligned with X axis in the initial position of the mesh (see Figure 2).
3. Nodes of a vane except the vane tip are elongated or shortened in the X direction. The nodes of the vane tip are just translated in the X direction.
4. A vane is rotated back by an angle θ to its actual location.
5. Nodes that lay on the rotor and cylinder surface are rotated by an angle β per time step. In the present work, $\beta = 0.1°$ was used.
6. The position of the internal nodes is calculated according to the following diffusion equation

$$\nabla \cdot \left(\Gamma_{disp} \nabla \delta\right) = 0 \tag{1}$$

where δ is the displacement of boundary nodes relative to the previous mesh locations and Γ_{disp} is the mesh stiffness, which determines the degree to which regions of nodes move together. For the present calculations, $\Gamma_{disp} = 1$ was used.

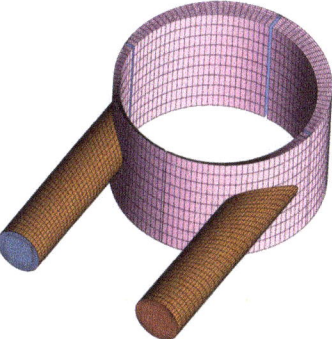

Figure 2. The initial position of the numerical mesh of the computational model at crank angle $\phi = 0°$. The working fluid flow was modelled as a compressible, turbulent, and single-phase Redlich–Kwong gas. The specific heat capacity of each tested medium was set as a function of temperature. The wall boundary was assumed adiabatic due to the brief time spent by a portion of the fluid inside the machine (see Table 1).

Table 1. Boundary conditions and model assumptions.

Parameter	Value	Remarks
Inlet temperature	40–90 °C	With steps of 10 degrees
Inlet pressure	Type of the working fluid and its saturation pressure at the inlet temperature	-
Condenser temperature	20 °C	-
Condenser pressure	Saturation pressure at 20 °C	depending on the working fluid
Time step	5.5×10^{-5} s	1 degree of revolution
Total time	0.1 s	5 full revolutions
Turbulence model	k-epsilon	Standard model
Wall heat transfer	Adiabatic	-
Vane-to-wall clearance	40 µm	constant

Parameters at the inlet to the expander were selected to be temperature and pressure. For each working fluid, the inlet temperature was varied in a range of 40–90 °C with step of 10 degrees. The inlet pressure was taken as a saturation pressure of the working fluid at the inlet temperature. All calculations have been performed for a fixed cold source temperature (condenser temperature) equal to 20 °C.

The rotor rotational speed was assumed constant at a nominal value of 3000 rpm for all considered cases. The temperature at the outlet of the discharge socket was assumed 20 °C for all the cases and the pressure was set to the saturation pressure given by the medium properties.

3. Results and Discussion

3.1. Analysis of the Influence of the Applied Working Fluid

Figure 3 shows pressure distribution in a working chamber versus crank angle ϕ for different working fluids and different inlet temperatures. Due to the saturation pressure differs for each working fluid the p-ϕ curves are different however they follow the same shape. With increasing the inlet temperature, the inlet pressure (i.e., saturation pressure) increases and the area under the curve in the p-ϕ diagram increases. The highest inlet pressure occurred for propane and the lowest for acetone. Variation of the curves is known from typical volumetric expanders except the strange recompression effect which is visible in the vicinity of ϕ = 200°. This behavior stems from the improper construction of the expander and it was described in detail in [3]. This effect is more visible for working fluids with a higher saturation pressure and it results in decrease of efficiency of the expander. What is more, the arrangement of control edges in the analyzed machine is chosen improperly, which is visible in pressure distributions depicted in Figure 3. The working fluid is held in the working chamber for too long a time, which results in the occurrence of recompression and negative pressure fluctuations.

For the sake of better understanding of how a working fluid influences the expander's performance, internal work was calculated as a difference between the inlet and the outlet enthalpy. The results of internal work are presented in Figure 4 in terms of pressure ratio (i.e., the ratio of the pressure at the inlet to the expander to the pressure at its outlet $\sigma = p_{in}/p_{out}$) for all working fluids used. Except the curve for acetone, all curves look similar in shape i.e., in the first phase they steeply increase and achieve a maximum in the range of $\sigma = 3$–4 (only propane for $\sigma \approx 2$). Next, with increasing σ gradual decrease of internal work is observed. However, in the case of acetone, the characteristic behaves contrarily and the minimum is visible for $\sigma \approx 7$. The results indicate that optimal pressure ratio for the rotary vane expander falls in the range of $\sigma = 3$–4. On the other side, the highest achievable internal work is obtained with propane for $\sigma \approx 2$. However, it could be better understand via Figure 5 where the internal efficiency of the expander versus pressure ratio is plotted. It is obvious now that using propane as a working fluid in the expander analyzed is very ineffective and features very low efficiency of the order of 5%. However, Figure 5 also confirms that the optimal pressure ratio is located in the range of σ = 3–4 for which the efficiency is in the range of 12–35%. The highest efficiency can be achieved for the acetone, however the pressure ratio has to be increased to about σ ≈ 5.

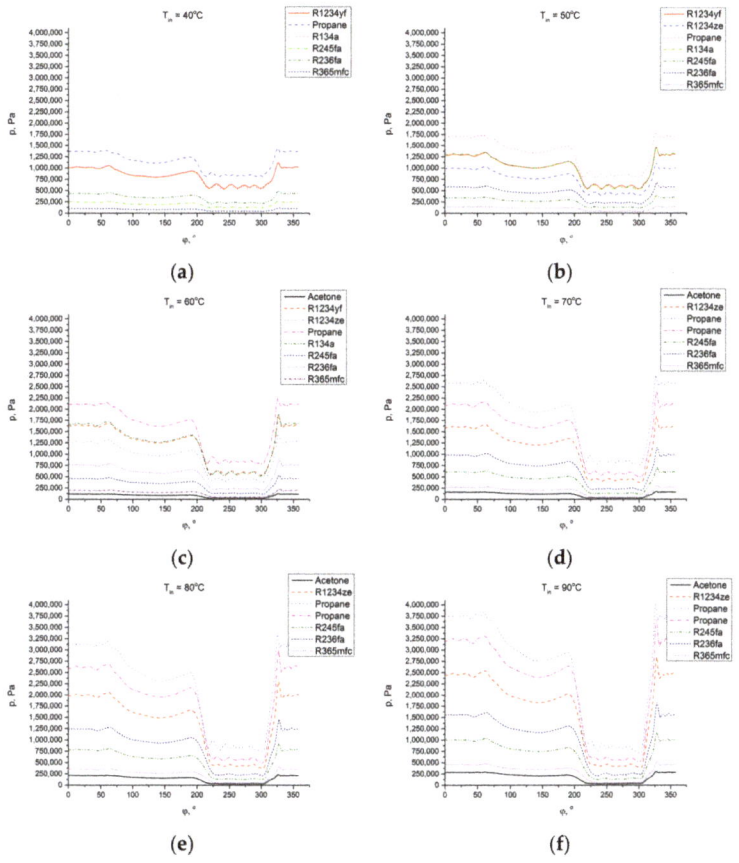

Figure 3. Pressure distribution against crank angle for different working fluids and inlet temperatures: (a) t_{in} = 40 °C; (b) t_{in} = 50 °C; (c) t_{in} = 60 °C; (d) t_{in} = 70 °C; (e) t_{in} = 80 °C; (f) t_{in} = 90 °C.

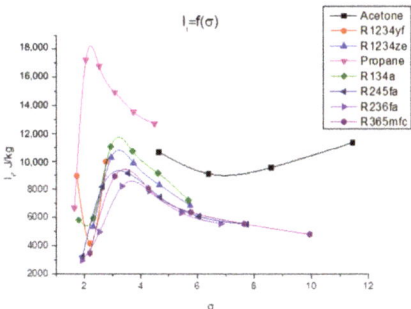

Figure 4. Internal work against pressure ratio for different working fluids.

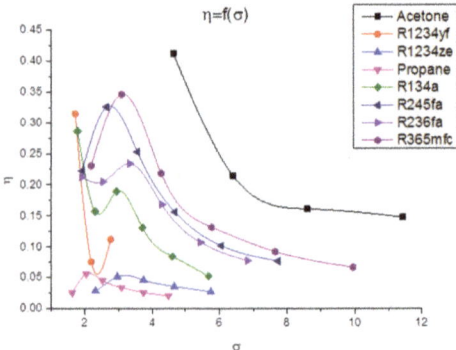

Figure 5. Efficiency against pressure ratio for different working fluids.

3.2. Analysis of the Influence of the Arrangement of the Steering Edges

Figure 6 shows the cross section of a multi-vane expander with marked main subassemblies and edges (referred to as machine steering edges) of the inlet (A, B) and the outlet port (C, D).

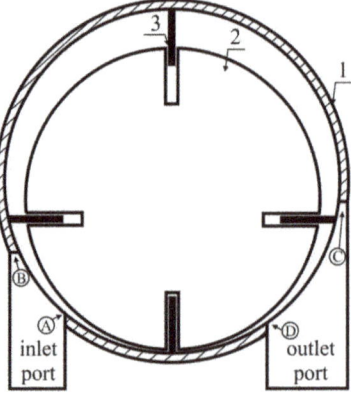

Figure 6. The cross section of the multi-vane expander at 0° crank angle. (1) cylinder; (2) rotor; (3) vane.

Steering edges act as valves that control the timing of the machine, therefore their mutual arrangement has an influence on the hydraulic losses, thermodynamic phenomena occurring in the working chambers of the machine and its performance [30,42]. Reference [42] provides the general guidelines for the proper selection of the mutual arrangement of the steering edges. According to these guidelines, the cylinder should be manufactured in such a way that edge B of the inlet port should be placed in the area where the working chamber reaches its maximum volume. Edge A of the inlet port should be placed in the area where the pressure of the working fluid contained in the working chamber before filling is equal to the pressure of the working fluid in the suction line. Edge C of the outlet port should be placed in the area where the pressure of the working fluid contained in the working chamber after expansion is equal to the pressure of the working fluid in the discharge line. Edge D of the outlet port should be placed in the area where the working chamber reaches its minimum volume.

The indicator diagram valid for the ideal multi-vane expander with the properly arranged steering edges is shown in Figure 7. It corresponds to a 1-2-3-4 broken curve. The visualization of the improper arrangement of the steering edges is also presented on this Figure, i.e., 1a and 1b visualize the process of the late start of suction; 1c and 1d visualize the process of early start of suction; 3a and 3b visualize early start of evacuation; and 3c and 3d visualize late start of evacuation. Improper construction of steering edges results in pressure fluctuations and decline in internal work. Consequently, thermal efficiency of a multi-vane expander decreases.

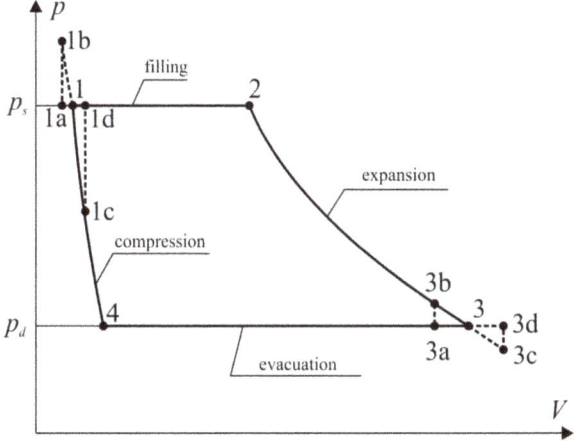

Figure 7. The multi-vane expander indicator diagram.

In the following paragraph, numerical analysis of the influence of arrangement of the steering edges is performed and the results of the numerical modelling are presented.

In order to numerically investigate the influence of machine steering edges on the expanders thermodynamic performance, modifications in the initial construction (referred to as 0° case) have been done. Figure 8 presents modified expander construction cases where machine steering edges are located in different places. Modification consisted in moving only the outlet port by a specified α angle while the inlet port remained in its initial position. The angle was varied by 22.5° and 0° corresponds to the initial position of the outlet port. The angles used were: 22.5°, 45°, 67.5°, and 90°. For these analysis purposes only, three working fluids were chosen, namely R123, R245fa, and R365mfc. Boundary conditions for each working fluid were set as stated in Table 2.

Table 2. Boundary conditions and model assumptions

Working Medium	t_{in}	p_{in}	σ	n
	°C	bar	-	rpm
R123		2.86	3.6	
R245fa	60 °C	4.63	4.3	3000
R365mfc		1.97	2.0	

The choice of R245fa and R365mfc working fluids is dictated by previous analysis which showed that from all analysed working fluids these two allows one to achieve the highest efficiency (see Figure 5). Additionally, R123 was analyzed as a reference working fluid.

Figure 9 shows an exemplary variation of the pressure inside the working chamber against crank angle for the case of R365mfc and five different angles of the outlet port position. As one can observe the crank angle that corresponds to onset of expansion is independent of the position angle α of the outlet port and remains close to $\phi \approx 90°$. For the initial case, i.e., $\alpha = 0°$, one can see the recompression effect that was mentioned earlier. When the position angle α of the outlet port increases, the recompression of the working fluid reduces and eventually ceases for $\alpha \geq 67.5°$. Simultaneously, the crank shaft angle at which working fluid is being evacuated from the working chamber decreases and the area under the curve in the p–V diagram increases, as is shown in Figure 10. Additionally, the pressure fluctuations described in previous paragraph resulted from improper location of the machine steering edges are lowered and eventually eliminated for $\alpha \geq 67.5°$.

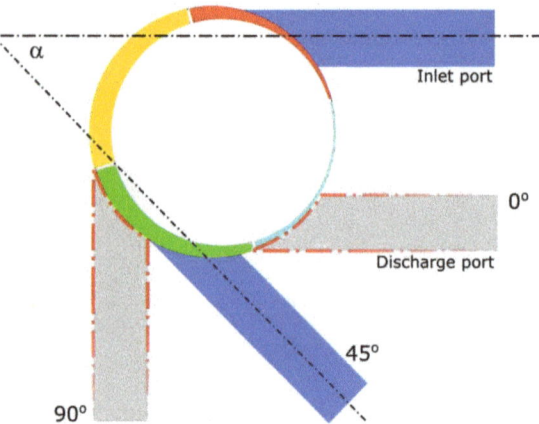

Figure 8. The modified multi-vane expander's construction with different locations of machine steering edges ($\alpha = 0°, 22.5°, 45°, 67.5°$, and $90°$).

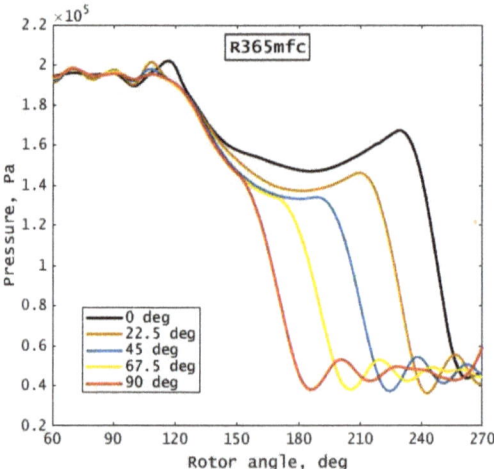

Figure 9. Pressure variation in a working chamber for R365mfc and different outlet port locations.

Elimination of the recompression effect and pressure fluctuations caused by improper design of the machine steering edges should result in better working conditions of the expander. To investigate this feature better, Figure 10 shows the p–V diagram of the expander for R365mfc as a working fluid and different location of the outlet port. For the initial case, i.e., $\alpha = 0°$ the recompression effect significantly decreases the area under p–V curve and the resulting internal work is also decreased. For the case of $\alpha = 45°$ the area under p–V curve is much bigger due to reduction of the recompression.

However, it is also observed that moving the outlet port by α angle results in displacement of point 4 (see Figure 7) towards larger volumes. This results in displacement of the onset of compression stage in the expander which now starts for higher volumes in p–V diagram and the evacuation stage is decreased. This behavior could be better presented in Figure 11, where the working fluid mass m_0 inside a working chamber is shown. It is visible that increasing α—i.e., moving the outlet port towards the inlet port—results in reducing the time of residence of the working fluid in the working chamber. Shorter time of residence of the working medium in the chamber advantageously limits the impact of the working fluid leakages between the adjacent working chambers. The mass flow into the chamber is distinctly higher for $\alpha < 45°$ due to adverse pressure difference between adjacent chambers. The high pressure at the chamber–outlet port interface caused by recompression results in increased leakage at the initial stage of expansion compared to the designs with $\alpha \geq 45°$. Also with increasing α angle, the mass of the working fluid inside the chamber before the filling phase is increased so the volumetric efficiency is lower. An earlier start of the evacuation also limits the energy losses related to recompression of the working fluid. On the other side, the area under p–V diagram is decreased and smaller internal work can be obtained. Nonetheless, according to Figure 10, for $\alpha = 45°$ net gain of the internal work is still positive. However, for the case of $\alpha = 90°$ the point 4 moves considerably on the right and the loss of the internal work is in this case larger than gain of the internal work due to elimination of the recompression phenomenon. It suggests the existence of an optimal steering edges arrangement for which the internal work is maximized.

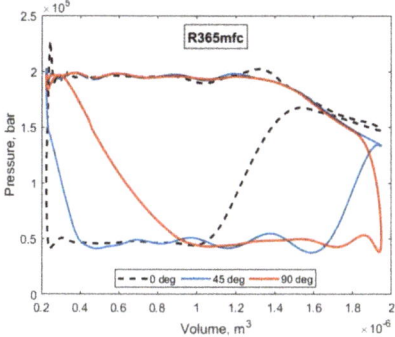

Figure 10. P–V diagram of the expander for R365mfc and different outlet port locations.

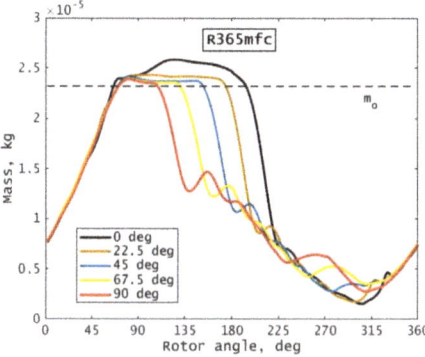

Figure 11. Mass of the working fluid in the working chamber of the expander versus rotor crank angle and different outlet port locations.

From the above considerations, one can infer that some optimal α angle should exist for which internal work and efficiency are the highest. To analyze it better, we calculated the output power and internal efficiency of the expander which are plotted in Figures 12 and 13, respectively. Figure 12 shows values of output power (bars) and outlet velocity (dashed lines) for all working fluids used and different discharge port positions. The highest output power is always obtained for R245fa working fluid. The displacement of the discharge port position results in almost double increase of power output and outlet velocity for R245fa, compared to R123 and R365mfc working fluids. It is also observed that in the range of α = 22.5–67.5° the maximum of output power exists. This feature is reflected in Figure 13 where thermal efficiency of the expander is plotted. The optimal parameters occur for $\alpha \approx 46°$. For R245fa and R365mfc working fluids, the maximum thermal efficiencies are 45% and 47%, respectively. On the other hand, for R123 the maximum reaches more than 55%. These results show that modification of the initial construction of the expander via applying proper machine steering edges position can significantly increase the thermal efficiency.

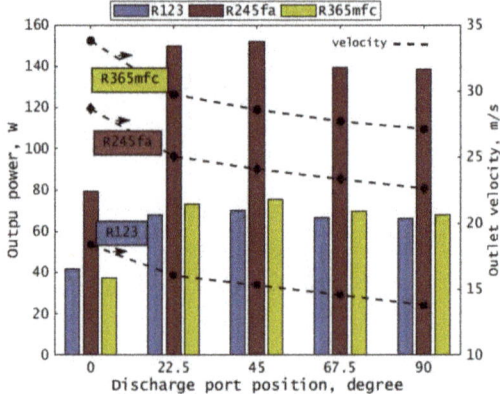

Figure 12. Output power and outlet velocity for different working fluids and different discharge port positions.

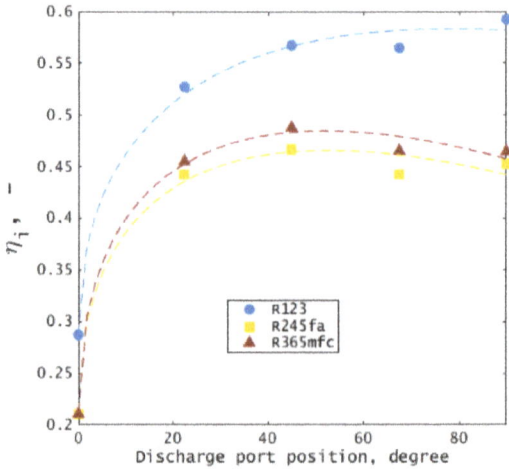

Figure 13. Variation of thermal efficiency of the expander for different working fluids and different outlet port positions.

Such high improvement of the expander's performance results from the proper arrangement of the machine steering edges which eliminates recompression and reduces pressure fluctuations. Additionally, the leakages between neighboring working chambers are limited as well. To understand it more precisely, Figure 14 shows the comparison of the size of the leakage from the inlet to the outlet port of the expander. For the initial construction of the expander (left figure) one can see that, for presented position of the vanes, the inlet and the outlet line are connected through the common working chamber. In this place, the highest pressure difference occurs and consequently it results in a large leaking mass flow. On the other side, for the improved construction (right figure) it is clearly visible that replacement of the outlet port positioning and thus change of the steering edges arrangement positively limits the impact of this phenomenon.

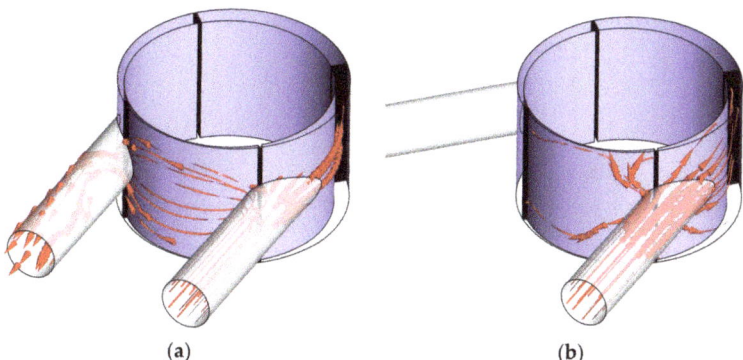

Figure 14. The visualization of leakage from the inlet to the outlet port of the expander: (**a**) present design; (**b**) improved design.

4. Summary and Conclusions

Three dimensional numerical analysis of a multi-vane expander was performed. The goal was to study the influence of the working fluid on the expander's performance. Additionally, an influence of machine steering edges arrangement on thermo-flow phenomena occurring in the expander was thoroughly studied. The numerical model was validated against the experiments that were carried out using a prototype of domestic ORC system equipped with a small multi-vane expander featuring a power of 300 W and utilizing R123 as a working fluid. The main aim of experimental investigations was to gather the input data needed for creating a numerical model of the expander and analysis of its operating conditions with a variety of working fluids.

The geometry of the expander used during the experiments was created and next divided into small elements that comprised numerical mesh. Governing equations of continuity, momentum, and energy together with equations of motion of the expander's elements were solved using commercial CFD tool. In order to analyze the influence of a working fluid on the expander's performance, different working fluids were selected and used i.e., R1234yf, R1234ze, R134a, R245fa, R236fa, R365mfc, acetone, and propane. During the numerical simulations, the value of temperature on the expander's inlet was varied similarly as during the experiments. It was found that for the expander analyzed and working fluids used, the optimal pressure ratio is in the range of $\sigma = 3$–4. Corresponding efficiency is in the range of 12–35% depending on working fluid. The numerical analysis showed that the best results were achieved for R245fa and R365mfc working fluids which featured the highest thermal efficiency.

On this basis R245fa, R365mfc, and R123 as a reference working fluid were used in the further step which was an attempt to optimize the expanders design in terms of efficiency. Based on calculated p–V diagrams, the influence of steering edges arrangement was recognized as a principal factor.

The numerical model has been modified to include variable discharge port position relative to the rotors eccentricity allowing to control the timing of working fluid evacuation. It was found that an optimal position of the machine steering edges exists for which the internal work and the thermal efficiency are maximum. For the analyzed expander's construction, the optimal position angle was $\alpha \approx 46°$. The obtained results show that the proper arrangement of the steering edges is crucial factor influencing the multi-vane expander performance and operating conditions.

Acknowledgments: Calculations have been carried out using resources provided by Wroclaw Center for Networking and Supercomputing (http://wcss.pl).

Author Contributions: Piotr Kolasiński conceived and designed the test stand and the experiments; Przemysław Błasiak and Józef Rak conceived and designed the numerical model; Piotr Kolasiński, Przemysław Błasiak and Józef Rak performed the experiments; Piotr Kolasiński, Przemysław Błasiak and Józef Rak analyzed the data; Piotr Kolasiński, Przemysław Błasiak and Józef Rak wrote the paper.

Conflicts of Interest: The authors declare no conflict of interest.

References

1. Hung, T.C.; Shai, T.Y.; Wang, S.K. A review of Organic Rankine Cycles (ORCs) for the Recovery of Low-Grade Waste Heat. *Energy* **1997**, *22*, 661–667. [CrossRef]
2. Kolasiński, P. The Influence of the Heat Source Temperature on the Multivane Expander Output Power in an Organic Rankine Cycle (ORC) System. *Energies* **2015**, *8*, 3351–3369. [CrossRef]
3. Mikielewicz, J.; Mikielewicz, D.; Ihnatowicz, E.; Kaczmarczyk, T.; Wajs, J.; Matysko, R.; Bykuć, S.; Rybiński, W. *Obiegi Termodynamiczne ORC Mikrosiłowni Domowej*; Institute of Fluid-flow Machinery Publishing: Gdańsk, Poland, 2014; ISBN 978-83-88237-19-5.
4. Stępień, R. *Wybrane Zagadnienia Projektowania Wielostopniowych Mikroturbin Osiowych*; Institute of Fluid-flow Machinery Publishing: Gdańsk, Poland, 2013; ISBN 978-83-88237-13-3.
5. Kozanecki, Z.; Kozanecka, D.; Klonowicz, P.; Łagodziński, J.; Gizelska, M.; Tkacz, E.; Miazga, M.; Kaczmarek, A. *Oil-Less Small Power Turbo-Machines (Bezolejowe Maszyny Przepływowe Małej Mocy)*; Institute of Fluid-flow Machinery Publishing: Gdańsk, Poland, 2014; ISBN 978-83-88237-27-0.
6. Mikielewicz, D.; Mikielewicz, J.; Ihnatowicz, E.; Muszyński, T.; Wajs, J.; Rybiński, W. *Wybrane Aspekty Projektowania i Badań Wymienników Ciepła Dla Obiegu ORC Mikrosiłowni Domowej*; Institute of Fluid-flow Machinery Publishing: Gdańsk, Poland, 2013.
7. Kaniecki, M.; Henke, A.; Krzemianowski, Z. *Agregaty Pompowe w Zastosowaniu do Obiegów ORC Mikrosiłowni Kogeneracyjnych na Czynnik Niskowrzący*; Institute of Fluid-flow Machinery Publishing: Gdańsk, Poland, 2013; ISBN 978-83-88237-14-0.
8. Kiciński, J.; Żywica, G. *Dynamika Mikroturbin Parowych*; Institute of Fluid-flow Machinery Publishing: Gdańsk, Poland, 2014.
9. Breńkacz, Ł.; Żywica, G.; Bogulicz, M. Selection of the Bearing System for a 1 kW ORC Microturbine. In Proceedings of the 31st Workshop on Turbomachinery, Dresden, Germany, 4–6 October 2017.
10. Żywica, G.; Kaczmarczyk, T.; Ihnatowicz, E.; Turzyński, T. Experimental investigation of the domestic CHP ORC system in transient operating conditions. *Energy Procedia* **2017**, *129*, 637–643. [CrossRef]
11. Kaczmarczyk, T.; Żywica, G.; Ihnatowicz, E. The impact of changes in the geometry of a radial microturbine stage on the efficiency of the micro CHP plant based on ORC. *Energy* **2017**, *137*, 530–543. [CrossRef]
12. Kaczmarczyk, T.; Żywica, G.; Ihnatowicz, E. Vibroacoustic diagnostics of a radial microturbine and a scroll expander operating in the organic Rankine cycle installation. *J. Vibroeng.* **2016**, *18*, 4130–4147. [CrossRef]
13. Kaczmarczyk, T.; Żywica, G.; Ihnatowicz, E. The Experimental Investigation of the Biomass-Fired ORC System with a Radial Microturbine. *Appl. Mech. Mater.* **2016**, *831*, 235–244. [CrossRef]
14. Kaczmarczyk, T.; Żywica, G.; Kiciński, J.; Ihnatowicz, E.; Turzyński, T.; Bykuć, S. Prototype of the Domestic CHP ORC System: Construction and Experimental Research. In Proceedings of the 3rd International Seminar on ORC Power Systems, Brussels, Belgium, 12–14 October 2015.
15. Kaczmarczyk, T.; Żywica, G.; Ihnatowicz, E. Experimental Investigation of a Radial Microturbine in Organic Rankine Cycle System with HFE7100 as Working Fluid. In Proceedings of the 3rd International Seminar on ORC Power Systems, Brussels, Belgium, 12–14 October 2015.

16. Klonowicz, P.; Witanowski, Ł.; Jędrzejewski, Ł.; Suchocki, T.; Lampart, P. A turbine based domestic micro ORC system. *Energy Procedia* **2017**, *129*, 923–930. [CrossRef]
17. Klonowicz, P.; Borsukiewicz-Gozdur, A.; Hanausek, P.; Kryłłowicz, W.; Brüggemann, D. Design and performance measurements of an organic vapour turbine. *Appl. Therm. Eng.* **2014**, *63*, 297–303. [CrossRef]
18. Klonowicz, P.; Heberle, F.; Preißinger, M.; Brüggemann, D. Significance of loss correlations in performance prediction of small scale, highly loaded turbine stages working in Organic Rankine Cycles. *Appl. Therm. Eng.* **2014**, *72*, 322–330. [CrossRef]
19. Lemort, V.; Guillaume, L.; Legros, A.; Declaye, S.; Quoilin, S. A Comparison of Piston, Screw and Scroll Expanders for Small Scale Rankine Cycle Systems. In Proceedings of the 3rd International Conference on Microgeneration and Related Technologies, Naples, Italy, 15–17 April 2013.
20. Song, P.P.; Zhuge, W.L.; Zhang, Y.J.; Zhang, L.; Duan, H. Unsteady Leakage Flow Through Axial Clearance of an ORC Scroll Expander. *Energy Procedia* **2017**, *129*, 355–362. [CrossRef]
21. Kosmadakis, G.; Mousmoulis, G.; Manolakos, D.; Anagnostopoulos, I.; Papadakis, G.; Papantonis, D. Development of Open-Drive Scroll Expander for an Organic Rankine Cycle (ORC) Engine and First Test Results. *Energy Procedia* **2017**, *129*, 371–378. [CrossRef]
22. Garg, P.; Karthik, G.M.; Kumar, P.; Kumar, P. Development of a generic tool to design scroll expanders for ORC applications. *Appl. Therm. Eng.* **2016**, *109*, 878–888. [CrossRef]
23. Gao, P.; Jiang, L.; Wang, L.W.; Wang, R.Z.; Song, F.P. Simulation and experiments on an ORC system with different scroll expanders based on energy and exergy analysis. *Appl. Therm. Eng.* **2015**, *75*, 880–888. [CrossRef]
24. Cao, Z.; Su, J.C.; Liu, Y.L.; Guo, B.; Chen, J.F. System Efficiency Comparison of Single-stage ORC with Twin-stage ORC Using Scroll Expanders. In Proceedings of the International Conference on Energy and Environment Engineering (ICEEE 2015), Nanjing, China, 11–12 April 2015.
25. Declaye, S.; Quoilin, S.; Guillaume, L.; Lemort, V. Experimental study on an open-drive scroll expander integrated into an ORC (Organic Rankine Cycle) system with R245fa as working fluid. *Energy* **2013**, *55*, 173–183. [CrossRef]
26. Kaczmarczyk, T.; Ihnatowicz, E.; Żywica, G.; Kiciński, J. Experimental investigation of the ORC system in a cogenerative domestic power plant with a scroll expanders. *Open Eng.* **2015**, *5*, 411–420. [CrossRef]
27. Fiaschi, D.; Secchi, R.; Galoppi, G.; Tempesti, D.; Ferrara, G.; Ferrari, L.; Karellas, S. Piston Expanders Technology as a Way to Recover Energy from the Expansion of Highly Wet Organic Refrigerants. In Proceedings of the ASME 9th International Conference on Energy Sustainability, San Diego, CA, USA, 28 June–02 July 2015.
28. Oudkerk, J.F.; Dickes, R.; Dumont, O.; Lemort, V. Experimental Performance of a Piston Expander in a Small-Scale Organic Rankine Cycle. In Proceedings of the 9th International Conference on Compressors and Their Systems, London, UK, 5–9 September 2015.
29. Norwood, Z.; Kammen, D.; Dibble, R. Testing of the Katrix rotary lobe expander for distributed concentrating solar combined heat and power systems. *Energy Sci. Eng.* **2014**, *2*, 61–76. [CrossRef]
30. Gnutek, Z.; Kolasiński, P. The application of rotary vane expanders in organic Rankine cycle systems—Thermodynamic description and experimental results. *J. Eng. Gas Turbines Power* **2013**, *135*, 61901. [CrossRef]
31. Kolasiński, P.; Błasiak, P.; Rak, J. Experimental investigation on multi-vane expander operating conditions in domestic CHP ORC system. *Energy Procedia* **2017**, *129*, 323–330. [CrossRef]
32. Murgia, S.; Valenti, G.; Colletta, D.; Costanzo, I.; Contaldi, G. Experimental investigation into an ORC-based low-grade energy recovery system equipped with sliding-vane expander using hot oil from an air compressor as thermal source. *Energy Procedia* **2017**, *129*, 339–346. [CrossRef]
33. Cipollone, R.; Contaldi, G.; Bianchi, G.; Murgia, S. Energy Recovery Using Sliding Vane Rotary Expanders. In Proceedings of the 8th International Conference on Compressors and Their Systems, London, UK, 9–10 September 2013.
34. Cipollone, R.; Bianchi, G.; Di Battista, D.; Contaldi, G.; Murgia, S. Mechanical energy recovery from low grade thermal energy sources. *Energy Procedia* **2014**, *45*, 121–130. [CrossRef]
35. Cipollone, R.; Bianchi, G.; Gualtieri, A.; Di Battista, D.; Mauriello, M.; Fatigati, F. Development of an Organic Rankine Cycle system for exhaust energy recovery in internal combustion engines. *J. Phys. Conf. Ser.* **2015**, *655*, 12015. [CrossRef]

36. Montenegro, G.; Della Torre, A.; Fiocco, M.; Onorati, A.; Benatzky, C.; Schlager, G. Evaluating the Performance of a Rotary Vane Expander for Small Scale Organic Rankine Cycles using CFD tools. In Proceedings of the ATI 2013—68th Conference of the Italian Thermal Machines Engineering Association, Bologna, Italy, 11–13 September 2013.
37. Kolasiński, P.; Błasiak, P.; Rak, J. Experimental and Numerical Analyses on the Rotary Vane Expander Operating Conditions in a Micro Organic Rankine Cycle System. *Energies* **2016**, *9*, 606. [CrossRef]
38. Vodicka, V.; Novotny, V.; Mascuch, J.; Kolovratnik, M. Impact of major leakages on characteristics of a rotary vane expander for ORC. *Energy Procedia* **2017**, *129*, 387–394. [CrossRef]
39. Bianchi, G.; Rane, S.; Kovacevic, A.; Cipollone, R.; Murgia, S.; Contaldi, G. Grid Generation Methodology and CFD Simulations in Sliding Vane Compressors and Expanders. In Proceedings of the 10th International Conference on Compressors and Their Systems, London, UK, 9–13 September 2017.
40. Rak, J.; Błasiak, P.; Kolasiński, P. Numerical modelling of multi-vane expander operating conditions in ORC systems. In Proceedings of the International Conference on Advances in Energy Systems And Environmental Engineering (ASEE17), Wrocław, Poland, 2–5 July 2017.
41. Bianchi, G.; Fatigati, F.; Murgia, S.; Cipollone, R.; Contaldi, G. Modeling and experimental activities on a small-scale sliding vane pump for ORC-based waste heat recovery applications. *Energy Procedia* **2016**, *101*, 1240–1247. [CrossRef]
42. Warczak, W. *Refrigerating Compressors*; WNT: Warsaw, Poland, 1987.

© 2018 by the authors. Licensee MDPI, Basel, Switzerland. This article is an open access article distributed under the terms and conditions of the Creative Commons Attribution (CC BY) license (http://creativecommons.org/licenses/by/4.0/).

Article

A Decentralized, Hybrid Photovoltaic-Solid Oxide Fuel Cell System for Application to a Commercial Building

Alexandros Arsalis [1,2,*] and George E. Georghiou [1,3]

[1] FOSS Research Centre for Sustainable Energy, University of Cyprus, Nicosia 1678, Cyprus; geg@ucy.ac.cy
[2] Department of Mechanical and Manufacturing Engineering, University of Cyprus, Nicosia 1678, Cyprus
[3] Department of Electrical and Computer Engineering, University of Cyprus, Nicosia 1678, Cyprus
* Correspondence: alexarsalis@gmail.com; Tel.: +357-22894396

Received: 5 November 2018; Accepted: 11 December 2018; Published: 16 December 2018

Abstract: New energy solutions are needed to decrease the currently high electricity costs from conventional electricity-only central power plants in Cyprus. A promising solution is a decentralized, hybrid photovoltaic-solid oxide fuel cell (PV-SOFC) system. In this study a decentralized, hybrid PV-SOFC system is investigated as a solution for useful energy supply to a commercial building (small hotel). An actual load profile and solar/weather data are fed to the system model to determine the thermoeconomic characteristics of the proposed system. The maximum power outputs for the PV and SOFC subsystems are 70 and 152 kWe, respectively. The average net electrical and total efficiencies for the SOFC subsystem are 0.303 and 0.700, respectively. Maximum net electrical and total efficiencies reach up to 0.375 and 0.756, respectively. The lifecycle cost for the system is 1.24 million USD, with a unit cost of electricity at 0.1057 USD/kWh. In comparison to the conventional case, the unit cost of electricity is about 50% lower, while the reduction in CO_2 emissions is about 36%. The proposed system is capable of power and heat generation at a lower cost, owing to the recent progress in both PV and fuel cell technologies, namely longer lifetime and lower specific cost.

Keywords: hybrid system; decentralized system; combined-heat-and-power; solid oxide fuel cells; photovoltaic; thermoeconomic modeling

---

1. Introduction

Efforts to increase energy efficiency have intensified over recent years due to the rapid increase of fossil fuel prices and also the need to decrease harmful emissions to the atmosphere [1]. Cogeneration allows the combination of various technologies to improve the fuel efficiency of electricity-only power plants or combined-heat-and-power (CHP) systems [2]. Earlier systems have included combined cycle power plants at large scale (10–100 MWe). In the resulting systems there exists a high level of complexity due to the increased number of parameters. Therefore, there is a need to apply advanced methodologies able to determine optimum solutions. However, this procedure becomes rigorous in case a number of parameters (e.g., thermodynamic, economic and environmental criteria) must be included [3]. Fuel cell technology has been proposed at the kW to the MW scale in a number of proposed systems. In lower temperature proton exchange membrane fuel cells (PEMFCs), CHP systems have been primarily applied at the kW scale, for smaller residential applications, where low-grade heat (recovered from the fuel cell exhaust) is usually adequate to cover residential load profiles, such as space heating and domestic hot water preparation [4,5]. These systems are sometimes operated jointly with vapor compression heat pumps to boost heat generation and/or to provide space cooling [6]. Such systems have been proposed for single-family households and in some cases for multi-unit residential applications [7,8]. Coupling of SOFCs with absorption chiller-heater units has also been proposed for larger scale, decentralized applications such as commercial buildings [9]. Although the resulting operational configurations have led to high system efficiencies (80% to 90%), their complexity resulted in high capital cost, which often dominates lifecycle cost.

Recent progress in SOFC technology includes important advances, such as higher lifetime, lower capital cost, higher electrical efficiency and simpler fuel processing requirements (in the case of natural gas-fueled systems—as compared to PEMFC technology) [10,11]. For small-scale residential applications, SOFC-based, natural gas-fueled micro-CHP systems have been proposed through thermoeconomic modeling and optimization techniques and improved operational strategies [12,13]. The application of effective optimization techniques, such as decomposition strategies, have been applied in some cases for the design/synthesis optimization of such systems [7,8]. For large-scale applications, the possibility of combining SOFC technology with heat engines, such as gas turbine cycles, has been thoroughly investigated since the early 2000s. In natural gas-fueled hybrid systems, where high temperature SOFC stacks have been integrated with gas turbine cycles, effort was placed on the increase of system efficiency to lower fuel consumption [14]. Due to the complexity of the proposed systems, the design/synthesis options are usually evaluated with advanced optimization techniques [15,16].

More recent research effort has focused on the possibility of combining fuel cell technology with renewable energy sources (RES). The combination of RES with fuel cell technology is a more environmentally friendly solution than decentralized hybrid photovoltaic (PV)-gas turbine systems, because in the latter case emissions are generated on-site, i.e., near the serviced buildings [17,18]. The deployment of PV units continues to increase because of significant cost reductions in addition to supportive policies, such as net-metering [19]. In such systems, excess generation of electricity from RES, e.g., via solar PV panels or wind turbines, can be converted to hydrogen through an electrolyzer unit [20], stored in a hydrogen storage tank, and then reconverted to electricity when renewable energy is unavailable [21]. The design of such systems for variable load has proven difficult and in most cases the proposed systems have considered grid-connected operation to allow import/export of electricity, while in other cases a constant load operation was followed [22,23]. A combination of RES with natural gas (or biogas)-fueled fuel cell units could allow a rapid deployment of these hybrid systems [24]. Currently the application of hybrid PV-SOFC systems seems more attractive for commercial buildings as the load demand closely matches the solar energy availability. The use of dynamic or quasi-steady state modeling is usually required to model the system as realistically as possible [25,26].

The objective of this research study is the thermoeconomic modeling of a decentralized, hybrid PV-SOFC system for application to a commercial building. The PV subsystem, the fuel cell stack, and the steam methane reformer (SMR) reactor components are modeled in detail to allow a realistic representation of their operation at both design and off-design conditions. In addition, a significant shortcoming of previous studies on hybrid RES-fuel cell systems is the fact that, in most cases, actual load profiles have not been considered. The omission of an actual load profile prohibits the extraction of realistic outcomes on the actual viability of such systems. The current study considers both solar/weather data and an actual load profile for a commercial building for the whole year. This approach leads to a more accurate determination of the thermoeconomic characteristics of the proposed system, allowing a direct comparison to conventional useful energy generation. The fuel processor (pre-reformer) is of the SMR type, since it is more efficient than other technologies (e.g., partial oxidation), allowing more efficient natural gas conversion to hydrogen [11]. The current research study investigates the economic competitiveness of the proposed system in comparison to conventional or alternative power generation. Four different cases are investigated and compared, namely: (A) Central power grid connection (conventional), (B) central power grid connection assisted with PV arrays, (C) non-grid connected SOFC system and (D) decentralized hybrid PV-SOFC system (proposed system). The outcomes of the research work are expected to reveal the possibility of combining and utilizing two highly advantageous technologies, i.e., PVs and solid oxide fuel cells, with an analysis beyond theoretical predictions. This is done with a detailed thermoeconomic modeling of the components, and further on with their overall integration in the system model. Moreover, through the development of a cost model, a complete thermoeconomic analysis is facilitated to lay out the characteristics of the proposed hybrid system.

2. System Configuration

The proposed system, shown in Figure 1, was designed to fully fulfill an actual load profile for a commercial building. It includes a natural gas-fueled SOFC subsystem and a solar PV subsystem. The system also includes DC/AC inverters to convert the DC current generated by the PV and the SOFC subsystems to AC electricity prior to distribution to the buildings. In the SOFC subsystem, natural gas (NG) is compressed in the fuel compressor and sulfur is removed with the desulfurizer. The NG is preheated through heat exchanger (HEx) HEx1 before entering the SMR. The endothermic process in the SMR requires external heating, which is generated by a catalytic combustor. The synthesis gas (syngas) at the SMR exit is fed to the fuel cell anode. Air drawn from the atmosphere is filtered and blown to HEx3 for preheating and then fed to the fuel cell cathode. The fuel cell reaction in the SOFC stack generates electricity and also a hot mixture at the fuel cell exit. The hot exhaust mixture is fed to the combustor, along with natural gas from the natural gas supply and air. The flue gas exiting the SMR is used to provide heat for the four heat exchangers (HEx1–HEx4). HEx2 is used to generate steam for the SMR. HEx4 is used to provide low-grade heat externally, i.e., to heat water from recovered heat and supply it to the hot water storage tank. Through the hot water storage tank, hot water is provided to the buildings. At the exit of HEx4, the exhaust flue gas is released to the atmosphere after separation of water through a water separator.

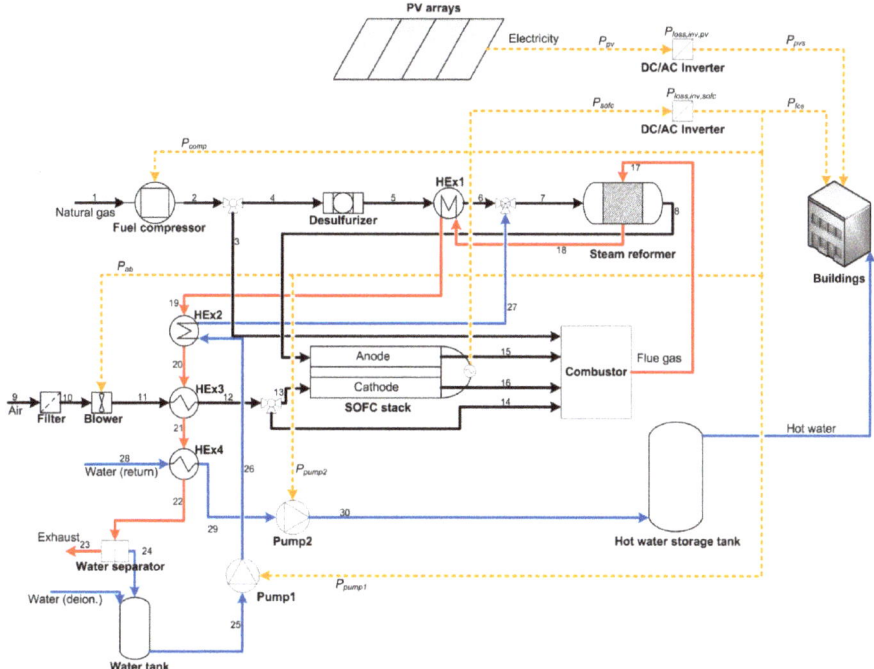

Figure 1. Schematic representation of the proposed hybrid photovoltaic-solid oxide fuel cell (PV-SOFC) system.

The main assumptions for the current study are the following:

1. The proposed system operates in complete autonomy, i.e., it is not connected to a central power grid (no import/export of electricity).
2. The maximum PV power output is set equal to the minimum electric load in the load profile to ensure no power is wasted. In turn, this value is used to size the SOFC subsystem. The system model is modeled in such a way to ensure that the proposed system is capable of completely covering the building load profile at all times, throughout the year.
3. Heat losses are considered in the three main components of the SOFC subsystem, namely: The SMR reactor, the SOFC stack, and the catalytic combustor. Additionally, pressure losses are considered in every component of the SOFC subsystem.
4. Additional heating (space heating and domestic hot water) is provided through natural gas-fired boilers, while space cooling is provided through electric vapor-compression heat pumps. This equipment is already available in the buildings and therefore its associated capital cost is not considered in the thermoeconomic modeling for this study.
5. The hourly solar and ambient temperature data used in the simulation of the PV subsystem are based on a Typical Meteorological Year—TMY2 for Nicosia, Cyprus [27].
6. The consumption data system is applied for a small hotel with load profile data extracted from [28]. The load profile includes the following loads (all in an electrical energy basis): Fans, interior equipment, interior lights, space cooling, space heating, and domestic hot water. The load profile is shown in Figure 2.

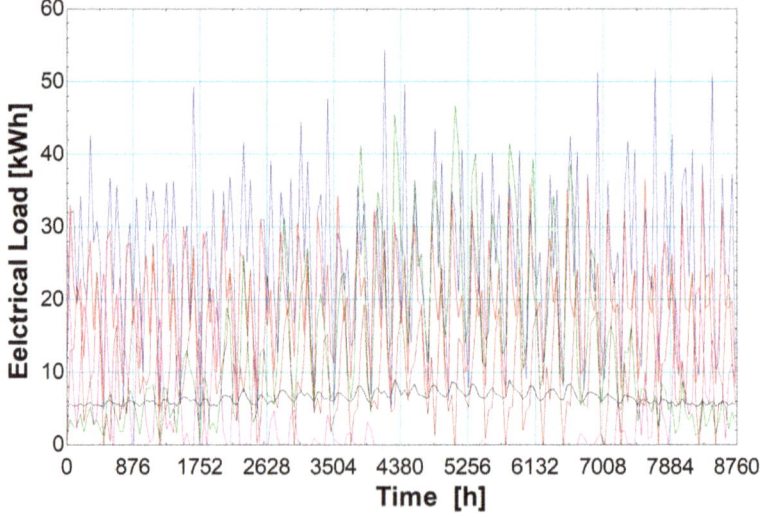

Figure 2. The load profile includes the following electrical loads (graph color in parenthesis): Fans (black), interior equipment (blue), interior lights (red), space cooling (green), space heating (purple), and domestic hot water (brown).

3. System Modeling

The modeling of the components of the proposed hybrid system was based on first principles to accurately represent the coupling and operation of the system as realistically as possible. After modeling each component, the components were coupled together to form the SOFC subsystem. Subsequently, simulation of the PV subsystem generates PV data for the simulation of the overall system model. Additionally, a cost model was developed for the economic analysis of the proposed system. It includes all necessary cost functions and inputs needed for the calculation of capital costs, fuel cost, lifecycle cost and unit cost of electricity. The modeling of the system was developed with the software Engineering Equation Solver (EES)—Professional version. Hourly simulation data were generated for a complete year, i.e., 8760 hourly segments.

3.1. Photovoltaic Subsystem

The PV subsystem was based on the Hay-Davies-Klucher-Reindl (HDKR) modeling methodology [29], i.e., the total incident solar radiation on a tilted surface is calculated with a consideration of both the ground-reflected and the beam effects:

$$I_T = (I_b + I_d \cdot A_i) \cdot R_b + I_d \cdot (1 - A_i) \cdot \left(\frac{1 + \cos\beta}{2}\right) \cdot \left[1 + f \cdot \sin^3\left(\frac{\beta}{2}\right)\right] + I \cdot \rho_g \cdot \left(\frac{1 - \cos\beta}{2}\right). \quad (1)$$

In the PV array, the temperature was calculated with the relation (the effect of wind speed is considered negligible):

$$\frac{T_c - T_{amb}}{T_{NOCT} - T_{amb,NOCT}} = \frac{I_T}{I_{T,refer}} \cdot \left(1 - \frac{\eta_{ref}}{0.9}\right). \quad (2)$$

The array's maximum power point efficiency is:

$$\eta_{mp} = \eta_{refer} \cdot \left(1 + \mu_{mp} \cdot (T_c - T_{amb,NOCT})\right). \quad (3)$$

The PV array's electricity output is:

$$P_{pv} = A_{pv,array} \cdot I_T \cdot \eta_{mp}. \quad (4)$$

3.2. SOFC Subsystem

The SOFC subsystem includes the fuel processing subsystem with the fuel pre-reformer (SMR reactor), four heat exchangers, SOFC stack and actuators. For the configuration shown in Figure 1, the inputs are given in Table 1. The fuel utilization factor was set at 0.92, and fuel cell temperature was set at 750 °C [10]. The temperature of fuel at the fuel preheater exit, the temperature of the reformate at the SMR reactor exit/anode inlet, and the temperature of the flue gas exiting the catalytic combustor were set at 450, 650 and 1005 °C, respectively [11]. HEx4 flue gas exit temperature was set at 55 °C because it must be at 25 °C above the dew point of the combustion product gases [10]. The steam-to-carbon ratio was set at 2.5, which although it is a relatively low value, the SOFC can treat CO as fuel [30], and therefore CO content does not need to be significantly reduced prior to anode inlet.

Table 1. System input parameters of the solid oxide fuel cell (SOFC) subsystem.

	Parameter Description	Value
U_f	Fuel utilization factor	0.92
A_{fc}	Fuel cell effective cross-sectional area	144 cm^2
n_{cells}	Total number of cells in fuel cell stacks	12,000
T_{fc}	Fuel cell operating temperature	750 °C
T_6	Fuel preheater exit temperature	450 °C
T_8	SMR reactor reformate exit temperature	650 °C
T_{13}	Cathode inlet temperature	650 °C
T_{17}	Combustor exit temperature	1005 °C
T_{22}	HEx4 flue gas exit temperature	55 °C
T_{25}	Water pump 1 inlet temperature	40 °C
T_{28}	Hot water storage tank return temperature	40 °C
T_{29}	Hot water storage tank supply temperature	65 °C
SC	Steam-to-carbon ratio	2.5

3.2.1. SMR Reactor

An SMR reactor configuration was assumed for the pre-reformer. The SMR reactor model is based on chemical equilibrium [10–12]. Two chemical reactions were included: SMR reaction (methane-steam), and water gas shift (WGS) reaction (carbon monoxide-steam) [31]. Since the SMR reaction is endothermic, heat must be supplied by an external source [32] (in this case from the catalytic combustor).

For the SMR reaction, i.e., $CH_4 + H_2O \rightleftharpoons CO_2 + 3H_2$, the overall change in Gibbs free energy is:

$$\Delta G_{smr} = -1 \cdot g_{CH_4} - 1 \cdot g_{H_2O} + 1 \cdot g_{CO} + 3 \cdot g_{H_2}, \tag{5}$$

$$arg1 = \left(\frac{-\Delta G_{smr}}{R \cdot T_{ref,out}} \right). \tag{6}$$

The equilibrium constant at the given temperature for the SMR reaction is:

$$K_{smr} = \exp(arg1). \tag{7}$$

For the WGS reaction, $CO + H_2O \rightleftharpoons CO_2 + H_2$, the overall change in Gibbs free energy is:

$$\Delta G_{wgs} = -1 \cdot g_{CO} - 1 \cdot g_{H_2O} + 1 \cdot g_{CO_2} + 1 \cdot g_{H_2}, \tag{8}$$

$$arg2 = \left(\frac{-\Delta G_{wgs}}{R \cdot T_{ref,out}} \right). \tag{9}$$

The equilibrium constant at the given temperature for the WGS reaction is:

$$K_{wgs} = \exp(arg2). \tag{10}$$

The molar flow output is defined as:

$$\dot{n}_{ref,out} = \dot{n}_{ref,in,CH_4} + \dot{n}_{ref,in,H_2O} + 2 \cdot X_{smr}. \tag{11}$$

The equilibrium constants for the aforementioned reactions are [33]:

$$K_{smr} \cdot y_{ref,out,CH_4} \cdot y_{ref,out,H_2O} = y_{ref,out,CO} \cdot y_{ref,out,H_2}^3 \cdot \left(\frac{P_{ref,out}}{P_{amb}} \right)^2, \tag{12}$$

$$K_{wgs} \cdot y_{ref,out,CO} \cdot y_{ref,out,H_2O} = y_{ref,out,CO_2} \cdot y_{ref,out,H_2}. \tag{13}$$

A molar flow rate balance for each species can be applied at the reformer inlet and outlet:

$$\dot{n}_{ref,out,CH_4} = \dot{n}_{ref,in,CH_4} - X_{smr}. \tag{14}$$

$$\dot{n}_{ref,out,H_2O} = \dot{n}_{ref,in,H_2O} - X_{smr} - X_{wgs}. \tag{15}$$

$$\dot{n}_{ref,out,CO} = X_{smr} - X_{wgs}. \tag{16}$$

$$\dot{n}_{ref,out,H_2} = 3 \cdot X_{smr} + X_{wgs}. \tag{17}$$

$$\dot{n}_{ref,out,CO_2} = X_{wgs}. \tag{18}$$

The flue gas temperature at exit is calculated through an energy balance in the reformer:

$$\dot{Q}_{heat,smr} + \dot{E}_{in,smr} = \dot{E}_{out,smr} + \dot{Q}_{loss,smr}. \tag{19}$$

3.2.2. SOFC Stack

The SOFC stack model includes both the fuel cell reaction and direct internal reforming processes. For the latter, the reforming process takes place at the surface of the catalysts (anode), where hydrogen gas is mixed with steam before entering the anode [34]. Internal reforming is identical to the SMR reactor modeling equations, described in Section 3.2.1.

For the fuel cell reaction, the open circuit voltage is modeled as follows [10–12]:

$$E_{ocv} = E_{ocv,0} + \frac{R \cdot T_{fc}}{2 \cdot F} \cdot \ln\left(\frac{y_{ano,H_2} \cdot p_{fc}}{p_{amb}} \cdot \sqrt{\frac{y_{cat,O_2} \cdot p_{fc}}{p_{amb}}} \middle/ \frac{y_{ano,H_2O} \cdot p_{fc}}{p_{amb}} \right). \tag{20}$$

The reversible voltage is:

$$E_{ocv,0} = \frac{-\Delta g^o{}_f}{2 \cdot F}. \tag{21}$$

The Gibbs free energy is:

$$\Delta g^o{}_f = 1 \cdot g_{H_2O} - 0.5 \cdot g_{O_2} - 1 \cdot g_{H_2}. \tag{22}$$

The activation losses are based on the Butler–Volmer equation, defined for the anode and cathode, respectively, to determine the current density:

$$i = i_{0,ano} \cdot \left(\exp\left(\alpha \cdot \frac{n_e \cdot F}{R \cdot T_{fc}} \cdot V_{act,ano} \right) - \exp\left(-(1-\alpha) \cdot \frac{n_e \cdot F}{R \cdot T_{fc}} \cdot V_{act,ano} \right) \right), \tag{23}$$

$$i = i_{0,cat} \cdot \left(\exp\left(\alpha \cdot \frac{n_e \cdot F}{R \cdot T_{fc}} \cdot V_{act,cat} \right) - \exp\left(-(1-\alpha) \cdot \frac{n_e \cdot F}{R \cdot T_{fc}} \cdot V_{act,cat} \right) \right), \tag{24}$$

where $i_{0,ano}$ and $i_{0,cat}$ are the exchange current densities for the anode and cathode, respectively:

$$i_{0,ano} = \gamma_{ano} \cdot \left(\frac{y_{ano,H_2} \cdot p_{fc}}{p_{amb}} \right) \cdot \left(\frac{y_{ano,H_2O} \cdot p_{fc}}{p_{amb}} \right)^{-0.5} \cdot \exp\left(-\frac{E_{act,ano}}{R \cdot T_{fc}} \right), \tag{25}$$

$$i_{0,cat} = \gamma_{cat} \cdot \left(\frac{y_{cat,O_2} \cdot p_{fc}}{p_{amb}} \right)^{0.25} \cdot \exp\left(-\frac{E_{act,cat}}{R \cdot T_{fc}} \right). \tag{26}$$

The activation overvoltage is determined as the sum of anode and cathode losses:

$$V_{act} = V_{act,ano} + V_{act,cat}. \tag{27}$$

Concentration losses are the gradual losses due to the reactant depletion in the catalyst layer, and they are defined as the difference between the Nernst potential at the catalyst layer and the bulk flow at both anode and cathode [35]. The limiting current densities for hydrogen and oxygen species are defined as follows, respectively:

$$i_{L,H_2} = 2 \cdot F \cdot C_{H_2,0} \cdot h_{m,H_2}, \tag{28}$$

$$i_{L,O_2} = 4 \cdot F \cdot C_{O_2,0} \cdot h_{m,O_2}, \tag{29}$$

where $C_{H_2,0}$ and $C_{O_2,0}$ are the concentration of species for hydrogen and oxygen, respectively.

The concentration losses are [35]:

$$V_{conc} = -\frac{R \cdot T_{fc}}{2 \cdot F} \cdot \ln\left(\left(1 - \frac{i}{i_{L,H_2}}\right) \cdot \left(1 - \frac{i}{i_{L,O_2}}\right)^{0.5} \right). \tag{30}$$

The Ohmic losses are defined as the product of current density and Ohmic resistance:

$$V_{ohm} = i \cdot R_i. \tag{31}$$

Based on the above definitions, the cell voltage can be defined as follows [34]:

$$V_{cell} = E_{ocv} - V_{act} - V_{conc} - V_{ohm}. \tag{32}$$

Fuel cell stack voltage, current, and power are defined as follows, respectively:

$$V_{fc} = V_{cell} \cdot n_{cells}, \tag{33}$$

$$I_{fc} = i \cdot A_{fc}, \tag{34}$$

$$\dot{P}_{sofc} = V_{fc} \cdot I_{fc}. \tag{35}$$

The molar flow rate of oxygen at the inlet of the cathode can be calculated through an energy balance:

$$\dot{Q}_{in,fc} = \dot{Q}_{out,fc} + \dot{Q}_{loss,fc} + \dot{P}_{sofc} \tag{36}$$

3.2.3. Auxiliary Components

The auxiliary components are the actuators (air blower, fuel compressors and two water pumps), the catalytic combustor and the four heat exchangers. The actuators were modeled using fundamental equations, while the catalytic combustor model was based on an energy balance of products and reactants. The modeling of the heat exchangers was based on the Logarithmic Mean Temperature Difference (LMTD) method.

3.3. Overall System

The proposed hybrid system includes two prime movers for the generation of electrical energy. Additionally, in the case of the SOFC subsystem, heat is generated and recovered for external use in the buildings to satisfy the heating loads. Therefore, an algorithm must be included in the code of the system model to relate fuel cell power output, PV power output and power demand. Additionally, since the system is non-grid connected, it must be ensured that no excess power is generated from the PV subsystem.

$$\begin{aligned}
&\text{If}(P_{load} > P_{pvs})\text{then} \\
&\quad P_{fcs} = P_{load} - P_{pvs} \\
&\quad P_{pv,exc} = 0 \\
&\text{Else} \\
&\quad \text{If}(P_{load} < P_{pvs})\text{then} \\
&\quad\quad P_{fcs} = 0 \\
&\quad\quad P_{pv,exc} = P_{pvs} - P_{load} \\
&\quad \text{Else} \\
&\quad\quad P_{fcs} = 0 \\
&\quad\quad P_{pv,exc} = 0 \\
&\quad \text{EndIf} \\
&\text{EndIf}
\end{aligned} \tag{37}$$

The inverter power losses for the PV subsystem are calculated as follows:

$$\dot{P}_{loss,inv,pv} = \dot{P}_{pv} \cdot (1 - \eta_{inv,pv}), \tag{38}$$

$$\dot{P}_{pvs} = \dot{P}_{pv} - \dot{P}_{loss,inv,pv}. \tag{39}$$

Similarly, for the SOFC subsystem:

$$\dot{P}_{loss,inv,sofc} = \dot{P}_{sofc} \cdot \left(1 - \eta_{inv,sofc}\right). \tag{40}$$

The net electrical power output for the SOFC subsystem is defined as follows:

$$\dot{P}_{fcs} = \dot{P}_{sofc} - \dot{P}_{loss,inv,sofc} - \dot{P}_{ab} - \dot{P}_{comp} - \dot{P}_{pump1} - \dot{P}_{pump2}. \tag{41}$$

The net electrical efficiency for the SOFC subsystem can be based on the lower heating value (LHV) or the higher heating value (HHV), respectively [36]:

$$\eta_{el,net,LHV} = \frac{\dot{P}_{fcs}}{\dot{E}_{fuel,LHV}}, \tag{42}$$

$$\eta_{el,net,HHV} = \frac{\dot{P}_{fcs}}{\dot{E}_{fuel,HHV}}. \tag{43}$$

The thermal efficiency is the ratio of recovered heat rate actually used to cover the building heating loads (fully or partly) to the chemical energy rate of the fuel consumed by the SOFC subsystem:

$$\eta_{th} = \frac{\dot{Q}_{th}}{\dot{E}_{fuel,LHV}} \tag{44}$$

The total SOFC subsystem efficiency is the sum of SOFC subsystem net electrical efficiency and thermal efficiency:

$$\eta_{fcs} = \eta_{el,net,LHV} + \eta_{th}. \tag{45}$$

The thermal-to-electric ratio is the ratio of recovered heat rate to net electrical power output:

$$TER = \frac{\dot{Q}_{th}}{\dot{P}_{fcs}}. \tag{46}$$

When the recovered heat from the SOFC subsystem is inadequate to cover the heating loads, additional heat must be generated externally:

$$P_{heat,net} = \max(0, (P_{heat} + P_{dhw}) - P_{th}). \tag{47}$$

The total load profile electrical energy requirement is the sum of electricity required to operate the fans, the interior lights, the interior equipment, the space cooling, and supplementary heating:

$$P_{load} = P_{fan} + P_{light} + P_{equip} + P_{cool} + P_{heat,net}. \tag{48}$$

3.4. Cost Model

A cost model was developed to determine the economic performance of the proposed hybrid system, based on the methodology found in [37]. The modeling equations are shown in Table 2, while the values of the constant parameters are given in Table 3. The specific cost of the PV array was set at 2.00 USD/W, which is based on approximate values given in [38]. The specific cost of the SOFC subsystem and the power subsystem were approximated from values given in [39], and they were set at 2.00 and 1.00 USD/W, respectively. The specific cost of the power subsystem included the two DC/AC inverters and the power conditioning components. The cost of fuel (i.e., natural gas) was set at 7.19 USD/MMBTU, which is the current cost in the European Union (EU) [40]. The lifetime was set at 20 years for the system (i.e., PV arrays and power subsystem) [19,37] and 5 years for the

SOFC subsystem [39], with a fuel cell operation factor set at 0.50, since the fuel cell is operated for approximately 50% of the time. The cost of the hot water storage tank was based on an approximation from values given in [41]. The values for the remaining parameters were taken from [37].

Table 2. Cost modeling equations for the proposed hybrid system.

	Variable Description (Unit)	Model Equation
C_{fc}	Cost of SOFC subsystem (USD)	$N_{lt,fc,adj} = N_{lt,fc}/z_{fc}\ n_{re} = N_{lt}/N_{lt,fc,adj}$ $C_{fc} = n_{re} \cdot c_{fcs} \cdot \dot{P}_{fc,max}$
c_{fuel}	Cost of fuel in the first year (USD)	$c_{fuel} = c_{\$MMBtu} \cdot (3.6/3.41)[\$/GJ] \cdot \left\|1 \times 10^{-9} \dfrac{\$/J}{\$/GJ}\right\|$
E_{py}	Annual fuel consumption (J)	$E_{py} = E_{fuel,in,yr} \cdot \left\|3600 \dfrac{J}{W\cdot h}\right\|$
c_{fy}	Annual cost of fuel (USD/year)	$c_{fy} = E_{py} \cdot c_{fuel}$
C_{pv}	Cost of PV arrays (USD)	$C_{pv} = c_{pvs} \cdot P_{pv,max}$
C_{inv}	Cost of power subsystem (USD)	$C_{inv} = c_{invs} \cdot \left(P_{pv,max} + \dot{P}_{fc,max}\right)$
C_{sys}	Total cost of system (USD)	$C_{sys} = C_{pv} + C_{fc} + C_{inv} + C_{hwst}$
C_{down}	Down payment (USD)	$C_{down} = (1 - f_{loan}) \cdot C_{sys}$
AP_n	Capital recovery factor (-)	$AP_n = \dfrac{r_n}{1-(1+r_n)^{-N_{LT}}}$ $r_1 = r_{mL} - i\ r_2 = r_{mL}\ r_3 = \dfrac{r_2-r_1}{0.01+r_1}\ r_4 = \dfrac{r_{mL}-r_e}{1+r_e}$
PA_n	Uniform series present worth factor (-)	$PA_n = (AP_n)^{-1}$
FP_n	Compound amount factor (-)	$FP_n = (1+r_n)^{-N_{LT}}$
PF_n	Present worth factor (-)	$PF_n = (FP_n)^{-1}$
C_{loan}	Cost of the loan (USD)	$C_{loan} = \dfrac{AP_1}{AP_2} \cdot f_{loan} \cdot C_{sys}$
D_{loan}	Tax deduction on the loan (USD)	$D_{loan} = t \cdot f_{loan} \cdot C_{sys} \left(\dfrac{AP_1}{AP_2} - \dfrac{AP_1 - r_1}{(1+r_1)\cdot AP_3}\right)$
C_{twc}	Total worth of capital (USD)	$C_{twc} = C_{down} + C_{loan} - D_{loan}$
D_{dep}	Linear depreciation of capital (USD)	$D_{dep} = t \cdot PA_2 \cdot (C_{sys}/N_{LT})$
D_{cred}	Tax credit (USD)	$D_{cred} = t_{cred} \cdot C_{sys}$
D_{salv}	Salvage worth (USD)	$D_{salv} = f_{salv} \cdot C_{sys} \cdot PF_2 \cdot (1-t_{salv})$
C_{prop}	Tax paid on property (USD)	$C_{prop} = f_{prop} \cdot C_{sys} \cdot t_{prop} \cdot (1-t)$
C_{omi}	Operation, maintenance and insurance cost (USD)	$C_{omi} = f_{omi} \cdot C_{sys} \cdot PA_2 \cdot (1-t)$
C_{tcf}	Total cost of fuel (USD)	$C_{tcf} = c_{fy} \cdot \left(\dfrac{1-t}{AP_4}\right)$
LCC	Life cycle cost (USD)	$LCC = C_{twc} + C_{prop} + C_{omi} + C_{tcf} - \left(D_{dep} + D_{cred} + D_{salv}\right)$
c_{el}	Unit cost of electricity (USD/kWh)	$P_{cs,life} = N_{lt} \cdot P_{load,yr}$ $c_{el} = LCC/P_{cs,life}$

Table 3. Parameters held constant in the cost model.

	Parameter Description	Value
c_{pvs}	Specific cost of PV arrays	2.00 USD/W
c_{fcs}	Specific cost of SOFC subsystem	2.00 USD/W
c_{invs}	Specific cost of power subsystem	1.00 USD/W
$c_{\$MMBtu}$	Cost of fuel (natural gas)	7.19 USD/MMBTU
N_{lt}	System lifetime	20 years
$N_{lt,fc}$	SOFC subsystem lifetime	5 years
z_{fc}	Fuel cell operation factor	0.50
C_{hwst}	Cost of hot water storage tank	5000 USD
r_e	Real fuel price escalation rate	0.10
i	Inflation rate	0.01
r_m	Market discount rate	0.06
r_{mL}	Market loan rate	0.05
f_{loan}	Fraction of the capital cost paid through a loan	0.80
t	Incremental income tax	0.40
t_{cred}	Tax credit	0.02
f_{salv}	Salvage fraction	0.10
t_{salv}	Salvage tax	0.20
f_{prop}	Property fraction	0.50
t_{prop}	Property tax	0.25
f_{omi}	Operation and maintenance fraction	0.01

4. Results and Discussion

In this section the system model is validated with available literature data. Then, the performance of the proposed hybrid system is presented in detail. Finally, the proposed hybrid system is compared with conventional and alternative system configurations to analyze and investigate its competitiveness in regard to key thermoeconomic parameters.

4.1. Validation

For the validation of the SOFC stack, relevant literature data from [42] were used. As shown in Figure 3, the literature data compare well against the simulation data generated by the system model, with only a small deviation in the results. The PV subsystem was validated in a previous publication by some of the authors [43].

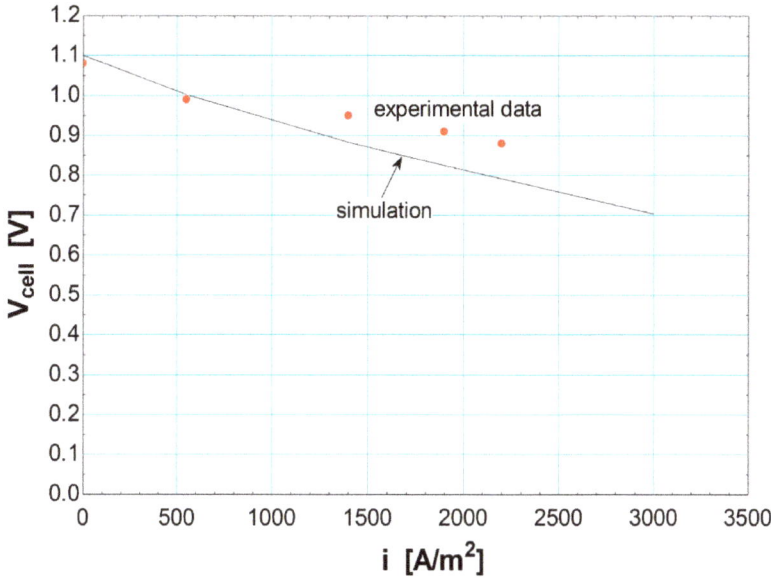

Figure 3. Validation of the modeled SOFC stack against literature data from [42].

4.2. Performance Characteristics of the Proposed Hybrid System

The proposed hybrid PV-SOFC system was sized in accordance with the requirements of the assumptions defined in Section 2. Based on these assumptions, the PV maximum power output is 70 kWe, while the SOFC maximum power output at full-load (i.e., design conditions) is 152 kWe. The average annual net electrical efficiency of the SOFC subsystem is 0.303, while total efficiency is 0.700. Maximum net electrical and total efficiencies can reach up to 0.375 and 0.756, respectively. In terms of annual useful energy generation, the electricity output (actual electricity delivered to the buildings) of the PV and SOFC subsystems is 135.9 and 451.2 MWh, respectively. The SOFC subsystem also provides 694.5 MWh of heating through heat recovery of the flue gas exhaust by the SOFC subsystem. This amount can almost completely cover the heating needs of the buildings, with only 7.2 MWh needed to be generated in addition. The current density at design conditions is 1228 A/m^2. A summary of the performance characteristics of the proposed system is given in Table 4. For an illustration of the performance of the system, Table 5 includes the values for the thermophysical parameters of the system at full load conditions for the SOFC subsystem.

Table 4. Performance characteristics of the proposed hybrid PV-SOFC system.

	Parameter Description	Value
$\dot{P}_{pv}$	PV maximum power output	70 kWe
$\dot{P}_{sofc}$	SOFC maximum power output	152 kWe
$\eta_{el,net,LHV}$	Net electrical efficiency of SOFC subsystem	0.200 (full-load) 0.303 (average) 0.375 (maximum) 0.659 (minimum)
η_{fcs}	Total efficiency of SOFC subsystem	0.700 (average) 0.756 (maximum) 0.8 (minimum)
TER	Thermal-to-electric ratio	1.3 (average) 2.5 (maximum)
P_{pvs}	Annual electricity output of PV subsystem	135.9 MWh
P_{fcs}	Annual electricity output of SOFC subsystem	451.2 MWh
$\dot{P}_{th}$	Annual heat recovery from SOFC subsystem	694.5 MWh
$P_{heat,net}$	Annual additional heat generation	7.2 MWh
P_{load}	Annual electricity load requirement	587.1 MWh
E_{fuel}	Annual fuel consumption	1610 MWh
$i_{fc,des}$	Current density at design conditions	1228 A/m^2
$\eta_{inv,pv}$	Inverter efficiency of PV subsystem	0.961 (average)
$\eta_{inv,sofc}$	Inverter efficiency of SOFC subsystem	0.970 (average)
U_1	Overall heat transfer coefficient of HEx1	90 W/m$^2\cdot$K
U_2	Overall heat transfer coefficient of HEx2	292 W/m$^2\cdot$K
U_3	Overall heat transfer coefficient of HEx3	4960 W/m$^2\cdot$K
U_4	Overall heat transfer coefficient of HEx4	8396 W/m$^2\cdot$K

Table 5. Values for the thermophysical parameters of the proposed PV-SOFC hybrid system at full load conditions for the SOFC subsystem.

Node	$\dot{n}$ (kg/s)	p (Pa)	T (K)	y_{CH_4}	y_{CO}	y_{CO_2}	y_{H_2}	y_{H_2O}	y_{N_2}	y_{O_2}
1	0.0008	130,000	298	0.000	0.000	0.000	0.000	0.000	0.000	0.000
2	0.0008	138,081	303	0.000	0.000	0.000	0.000	0.000	0.000	0.000
3	0.0003	119,800	303	0.000	0.000	0.000	0.000	0.000	0.000	0.000
4	0.0005	138,081	303	0.000	0.000	0.000	0.000	0.000	0.000	0.000
5	0.0005	128,081	303	0.000	0.000	0.000	0.000	0.000	0.000	0.000
6	0.0005	126,800	723	0.000	0.000	0.000	0.000	0.000	0.000	0.000
7	0.0016	126,800	723	0.286	0.000	0.000	0.000	0.714	0.000	0.000
8	0.0016	121,800	923	0.040	0.080	0.077	0.546	0.257	0.000	0.000
9	0.0294	101,325	298	0.000	0.000	0.000	0.000	0.000	0.000	0.000
10	0.0294	101,325	298	0.000	0.000	0.000	0.000	0.000	0.000	0.000
11	0.0294	124,040	320	0.000	0.000	0.000	0.000	0.000	0.000	0.000
12	0.0294	122,800	923	0.000	0.000	0.000	0.000	0.000	0.000	0.000
13	0.0294	122,800	923	0.000	0.000	0.000	0.000	0.000	0.000	0.000
14	0.0000	119,800	923	0.000	0.000	0.000	0.000	0.000	0.000	0.000
15	0.0017	119,800	1023	0.002	0.039	0.142	0.055	0.763	0.000	0.000
16	0.0288	119,800	1023	0.000	0.000	0.000	0.000	0.000	0.805	0.195
17	0.0308	117,800	1278	0.000	0.000	0.020	0.000	0.066	0.753	0.160
18	0.0308	115,300	1192	0.000	0.000	0.020	0.000	0.066	0.753	0.160
19	0.0308	114,147	1183	0.000	0.000	0.020	0.000	0.066	0.753	0.160
20	0.0308	113,006	1121	0.000	0.000	0.020	0.000	0.066	0.753	0.160
21	0.0308	111,875	586	0.000	0.000	0.020	0.000	0.066	0.753	0.160
22	0.0308	110,757	328	0.000	0.000	0.020	0.000	0.066	0.753	0.160
23	0.0287	110,757	328	0.000	0.000	0.022	0.000	0.000	0.807	0.171
24	0.0020	110,757	328	0.000	0.000	0.000	0.000	0.000	0.000	0.000
25	0.0012	101,325	313	0.000	0.000	0.000	0.000	0.000	0.000	0.000
26	0.0012	128,081	313	0.000	0.000	0.000	0.000	0.000	0.000	0.000
27	0.0012	126,800	723	0.000	0.000	0.000	0.000	0.000	0.000	0.000
28	0.1743	110,000	313	0.000	0.000	0.000	0.000	0.000	0.000	0.000
29	0.1743	108,900	338	0.000	0.000	0.000	0.000	0.000	0.000	0.000
30	0.1743	120,000	338	0.000	0.000	0.000	0.000	0.000	0.000	0.000

4.3. Thermoeconomic Analysis of the Proposed Hybrid System

4.3.1. System Cost Analysis

A cost analysis of the proposed hybrid system is given in Table 6. In terms of capital cost, the highest cost is allocated for the purchase of the SOFC subsystem at 607,540 USD, while the cost for the PV arrays is 140,132 USD. The cost of the power subsystem is also significant at 221,951 USD, which means that it constitutes about $\frac{1}{4}$ of the total system cost. The total cost of fuel for the operation of the system during its lifetime is estimated at 891,735 USD. Although natural gas prices are constantly fluctuating, it is not expected that this cost estimation will be significantly altered in the near future for the EU market, based on a statistical analysis of the prices for the last 10 years [40]. The lifecycle cost for the system is 1,241,369 USD, with a unit cost of electricity at 0.1057 USD/kWh.

Table 6. Cost analysis of the proposed hybrid PV-SOFC system.

	Output Parameter Description	Value
c_{fy}	Annual cost of fuel	43,995 USD/year
C_{pv}	Cost of PV arrays	140,132 USD
C_{fc}	Cost of SOFC subsystem	607,540 USD
C_{inv}	Cost of power subsystem	221,951 USD
C_{sys}	Total cost of the system	974,623 USD
C_{down}	Down payment	194,925 USD
C_{loan}	Cost of the loan	714,977 USD
D_{loan}	Tax deduction on the loan	236,951 USD
C_{twc}	Total worth of capital	672,950 USD
D_{dep}	Depreciation of capital	242,919 USD
D_{cred}	Tax credit	19,492 USD
D_{salv}	Salvage worth	206,877 USD
C_{prop}	Tax paid on property	73,097 USD
C_{omi}	Cost of operation, maintenance and insurance	72,876 USD
C_{tcf}	Total cost of fuel	891,735 USD
LCC	Lifecycle cost	1,241,369 USD
c_{el}	Unit cost of electricity	0.1057 USD/kWh

4.3.2. Case Study: Comparison with Conventional and Other System Configurations

For a quantitative investigation of the possible merits of the proposed hybrid PV-SOFC system, four different case configurations were analyzed, in terms of thermoeconomic performance: (A) Power supply from a central power grid (conventional case), (B) power supply from a central power grid assisted with an on-site PV system, (C) power (and heat) supply from a decentralized SOFC system, and (D) power (and heat) supply from the proposed hybrid PV-SOFC system. A schematic representation of the four cases is given in Figure 4. The four cases can be compared in terms of two parameters: Unit cost of electricity and CO_2 emissions. The results from this comparison are shown in Figures 5 and 6, respectively. As observed, the proposed system outperforms all other configurations, in terms of both the unit cost of electricity and CO_2 emissions. In particular, in comparison to case A, the unit cost of electricity is about 50% lower (0.2128 vs. 0.1057 USD/kWh), while the reduction in CO_2 emissions is about 36% (673 vs. 428 g(CO_2)/kWh).

In comparison to cases B and C, the additional capital cost for purchasing the SOFC subsystem and the PV subsystem, respectively, is well reasoned by the reduction in fuel consumption, hence on the unit cost of electricity (0.1700 USD/kWh (case B) and 0.1265 USD/kWh (case C)). Similarly, in terms of CO_2 emissions, the proposed system manages to significantly reduce emissions. For case B, power generation remains heavily dependent on inefficient central power grid supply, and therefore CO_2 emission generation remains high. For case C, CO_2 emissions are even higher than case B, because power (and heat) generation is completely dependent on the SOFC system. On an annual basis, the fuel consumption is 154,530 kg of natural gas for case C, compared to a reduced consumption of 115,848 kg for the proposed system in case D. In terms of lifecycle cost, for case C this is 1,468,209 USD, i.e., 226,840 USD higher than the equivalent cost for the proposed system in case D.

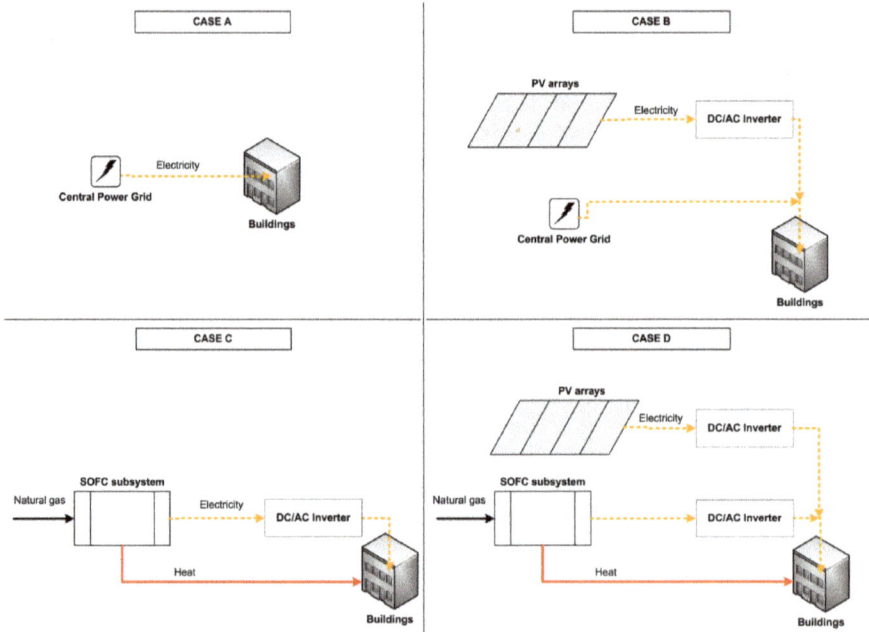

Figure 4. Schematic representation of the four considered cases: (**A**) Power supply from a central power grid (conventional case), (**B**) power supply from a central power grid assisted with an on-site PV system, (**C**) power (and heat) supply from a decentralized SOFC system, and (**D**) power (and heat) supply from the proposed hybrid PV-SOFC system.

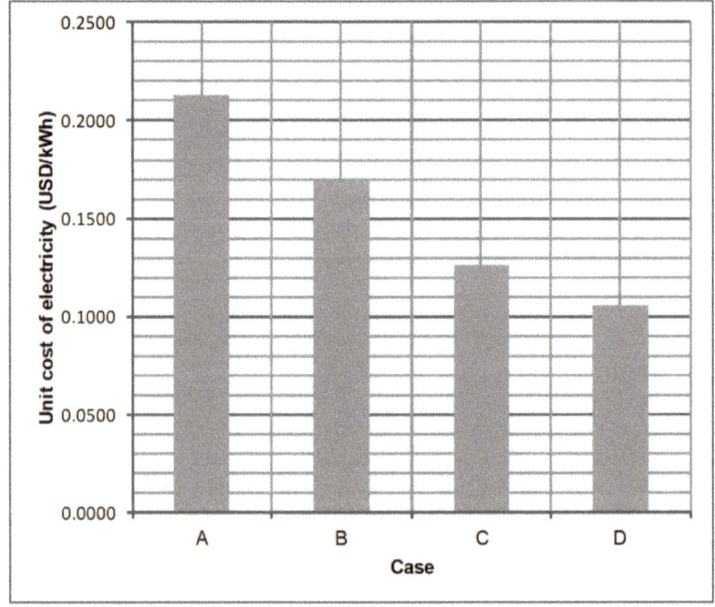

Figure 5. Comparison of the four cases under study in terms of the unit cost of electricity (in USD/kWh).

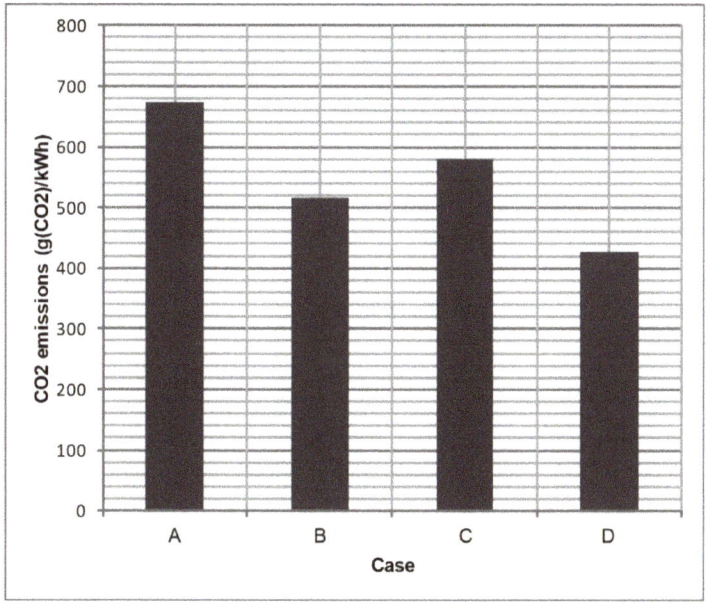

Figure 6. Comparison of the four cases under study in terms of CO_2 emissions (in g (CO_2)/kWh).

5. Conclusions

In this study a decentralized, hybrid PV-SOFC system is proposed for the fulfillment of a load profile for a commercial building (small hotel) in Cyprus. The system components are modeled in detail to allow a realistic simulation of the operation of the system. An actual load profile and solar/weather data are fed to the system model to determine the thermoeconomic characteristics of the proposed system. The system is sized based on the requirements of the load profile, with maximum power outputs for the PV and SOFC subsystems at 70 and 152 kWe, respectively. The system operates efficiently throughout the whole year for a transient load profile. The average net electrical and total efficiencies for the SOFC subsystem are 0.303 and 0.700, respectively. Maximum net electrical and total efficiencies reach up to 0.375 and 0.756, respectively. The total contribution of the two subsystems on a yearly basis for the fulfillment of the load profile is at 135.9 and 451.2 MWh for the PV and the SOFC subsystems, respectively. Application of the proposed hybrid system is favored over conventional power generation with electricity-only central power stations for technical and economic reasons. The proposed system can operate more efficiently in terms of net electrical efficiency (especially at part-load operation over a heat engine-based power generator), and, more importantly, it can take advantage of the heat recovery capability of the SOFC subsystem. Additionally, fuel consumption is reduced significantly, primarily because of the integration of the PV subsystem, and also due to the elimination of transmission and distribution losses.

The cost analysis of the proposed system shows that in terms of capital cost, the highest cost is for the purchase of the SOFC subsystem (607,540 USD), while the cost for the PV arrays is 140,132 USD. The cost of the power subsystem, which is usually underestimated, is also significant at 221,951 USD. The total cost of fuel for the operation of the system during its lifetime is estimated at 891,735 USD. Although natural gas prices are constantly fluctuating, it is not expected that this cost estimation will be significantly altered in the near future for the EU market, based on a statistical review of the prices for the last 10 years [33]. The lifecycle cost for the system is 1,241,369 USD, with a unit cost of electricity at 0.1057 USD/kWh. The proposed system outperforms conventional and other system configurations, in terms of both the unit cost of electricity and CO_2 emissions. In comparison to the conventional case, the unit cost of electricity is about 50% lower (0.2128 vs. 0.1057 USD/kWh), while the reduction in CO_2 emissions is about 36% (673 vs. 428 g(CO_2)/kWh). The additional capital cost for purchasing the PV and the SOFC subsystems is well reasoned by the reduction in fuel consumption, hence on the unit cost of electricity. Similarly, in terms of CO_2 emissions, the proposed system manages to significantly reduce emissions, because power generation is independent of the inefficient central power grid supply. Additionally, the integration of the PV subsystem allows a significant reduction in power generation from the SOFC subsystem during solar energy availability.

Author Contributions: Conceptualization, A.A.; Data curation, A.A.; Formal analysis, A.A.; Investigation, A.A.; Methodology, A.A.; Project administration, A.A.; Resources, G.E.G.; Software, A.A.; Supervision, G.E.G.; Validation, A.A.; Visualization, A.A.; Writing–original draft, A.A.; Writing–review & editing, A.A.

Funding: This research received no external funding.

Conflicts of Interest: The authors declare no conflict of interest.

Nomenclature and Units

A	Activation area (m^2)
A_i	Anisotropy index (-)
C	Concentration of species (kmol m^{-3})
E_{ocv}	Open circuit voltage (V)
$\dot{E}$	Energy rate (W)
f	Solar fraction (-)
F	Faraday's constant (Coulomb mol^{-1})
g	Specific Gibbs free energy (J $kmol^{-1}$)
h_m	Average diffusivity (m s^{-1})
i	Current density (A m^{-2})
I	Hourly irradiation (MJ m^{-2}), Current (A)
I_b	Beam radiation (MJ m^{-2})
I_d	Diffuse radiation (MJ m^{-2})
I_T	Total incident solar radiation (MJ m^{-2})
K	Reaction equilibrium constant (-)
$\dot{n}$	Molar flow rate (kmol s^{-1})
n_{cells}	Number of cells (-)
n_e	Number of electrons transferred per H_2 molecule reacted (-)
p	Pressure (bar, Pa)
P	Energy (kWh, MWh)
$\dot{P}$	Power (W)
$\dot{Q}$	Heat rate (W)
R	Universal gas constant (J mol^{-1} K^{-1})
R_b	Ratio of beam radiation (-)
R_i	Ohmic resistance (Ω m^2)
SC	Steam-to-carbon ratio (-)
T	Temperature (°C, K)
TER	Thermal-to-electric ratio (-)
U_f	Fuel utilization factor (-)
V	Voltage (V)
X	Conversion molar flow rate (kmol s^{-1})
y	Mole fraction (-)
Greek symbols	
α	Charge transfer coefficient (-)
β	PV tilt angle (degrees)
γ	Activity coefficient (A m^{-2})
Δg_f^0	Gibbs free energy (J $kmol^{-1}$)
ΔG	Overall change in Gibbs free energy (J $kmol^{-1}$)

Greek symbols
η	Efficiency (-)
μ_{mp}	Maximum power point efficiency temperature coefficient (-)
ρ_g	Ground reflectance (-)

Subscripts/Superscripts
0	Theoretical (ideal) value
ab	Air blower
act	Activation
amb	Ambient conditions
ano	Fuel cell anode
$array$	Array
c	PV array
cat	Fuel cell cathode
$cell$	Cell
$comp$	Fuel compressor
$conc$	Concentration
$cool$	Space cooling
dhw	Domestic hot water
el	Electrical
$equip$	Interior equipment
exc	Excess
fan	Fans
fc	Fuel cell
fcs	Fuel cell subsystem
$fuel$	Fuel
$heat$	Heat
HHV	Higher heating value
in	Inlet flow
inv	Inverter
L	Limiting
LHV	Lower heating value
$light$	Interior lights
$load$	Load
$loss$	Loss
mp	Maximum point
net	Net value
$NOCT$	Nominal operating cell temperature
ohm	Ohmic
out	Exit flow
$pump$	Water pump
pv	Photovoltaic
pvs	Photovoltaic subsystem
ref	Reformer
$refer$	Reference state
smr	Steam methane reformer
$sofc$	Solid oxide fuel cell
th	Recovered heat from fuel cell
wgs	Water gas shift

Abbreviations
CHP	Combined-heat-and-power
EES	Engineering Equation Solver
EU	European Union
HDKR	Hay-Davies-Klucher-Reindl

Abbreviations

HEx	Heat exchanger
HHV	Higher heating value
LHV	Lower heating value
LMTD	Logarithmic Mean Temperature Difference
PEMFC	Proton exchange membrane fuel cell
PV	Photovoltaic
RES	Renewable energy sources
SMR	Steam methane reformer
SOFC	Solid oxide fuel cell
WGS	Water gas shift

References

1. Moné, C.D.; Chau, D.S.; Phelan, P.E. Economic feasibility of combined heat and power and absorption refrigeration with commercially available gas turbines. *Energy Convers. Manag.* **2001**, *42*, 1559–1573. [CrossRef]
2. Popli, S.; Rodgers, P.; Eveloy, V. Trigeneration scheme for energy efficiency enhancement in a natural gas processing plant through turbine exhaust gas waste heat utilization. *Appl. Energy* **2012**, *93*, 624–636. [CrossRef]
3. Pelster, S.; Favrat, D.; von Spakovsky, M.R. The Thermoeconomic and Environomic Modeling and Optimization of the Synthesis, Design, and Operation of Combined Cycles With Advanced Options. *J. Eng. Gas Turbines Power* **2001**, *123*, 717–726. [CrossRef]
4. Mamaghani, A.H.; Najafi, B.; Casalegno, A.; Rinaldi, F. Predictive modelling and adaptive long-term performance optimization of an HT-PEM fuel cell based micro combined heat and power (CHP) plant. *Appl. Energy* **2017**, *192*, 519–529. [CrossRef]
5. Arsalis, A.; Kær, S.K.; Nielsen, M.P. Modeling and optimization of a heat-pump-assisted high temperature proton exchange membrane fuel cell micro-combined-heat-and-power system for residential applications. *Appl. Energy* **2015**, *147*, 569–581. [CrossRef]
6. Gunes, M.B.; Ellis, M.W. Evaluation of Energy, Environmental, and Economic Characteristics of Fuel Cell Combined Heat and Power Systems for Residential Applications. *J. Energy Resour. Technol.* **2003**, *125*, 208–220. [CrossRef]
7. Oyarzábal, B.; von Spakovsky, M.R.; Ellis, M.W. Optimal Synthesis/Design of a Pem Fuel Cell Cogeneration System for Multi-Unit Residential Applications–Application of a Decomposition Strategy. *J. Energy Resour. Technol.* **2004**, *126*, 30–39. [CrossRef]
8. Oyarzábal, B.; Ellis, M.W.; von Spakovsky, M.R. Development of Thermodynamic, Geometric, and Economic Models for Use in the Optimal Synthesis/Design of a PEM Fuel Cell Cogeneration System for Multi-Unit Residential Applications. *J. Energy Resour. Technol.* **2004**, *126*, 21–29. [CrossRef]
9. Weber, C.; Koyama, M.; Kraines, S. CO_2-emissions reduction potential and costs of a decentralized energy system for providing electricity, cooling and heating in an office-building in Tokyo. *Energy* **2006**, *31*, 2705–2725. [CrossRef]
10. Braun, R.J.; Klein, S.A.; Reindl, D.T. Evaluation of system configurations for solid oxide fuel cell-based micro-combined heat and power generators in residential applications. *J. Power Sources* **2006**, *158*, 1290–1305. [CrossRef]
11. Liso, V.; Olesen, A.C.; Nielsen, M.P.; Kær, S.K. Performance comparison between partial oxidation and methane steam reforming processes for solid oxide fuel cell (SOFC) micro combined heat and power (CHP) system. *Energy* **2011**, *36*, 4216–4226. [CrossRef]
12. Braun, R.J. Techno-economic optimal design of solid oxide fuel cell systems for micro-combined heat and power applications in the U.S. *J. Fuel Cell Sci. Technol.* **2010**, *7*, 310181–3101815. [CrossRef]
13. Antonucci, V.; Branchini, L.; Brunaccini, G.; De Pascale, A.; Ferraro, M.; Melino, F.; Orlandini, V.; Sergi, F. Thermal integration of a SOFC power generator and a Na – NiCl 2 battery for CHP domestic application. *Appl. Energy* **2017**, *185*, 1256–1267. [CrossRef]

14. Kurz, R. Parameter Optimization on Combined Gas Turbine-Fuel Cell Power Plants. *J. Fuel Cell Sci. Technol.* **2005**, *2*, 268–273. [CrossRef]
15. Calise, F.; Palombo, A.; Vanoli, L. Design and partial load exergy analysis of hybrid SOFC-GT power plant. *J. Power Sources* **2006**, *158*, 225–244. [CrossRef]
16. Calise, F.; Dentice d'Accadia, M.; Vanoli, L.; von Spakovsky, M.R. Single-level optimization of a hybrid SOFC-GT power plant. *J. Power Sources* **2006**, *159*, 1169–1185. [CrossRef]
17. Comodi, G.; Renzi, M.; Cioccolanti, L.; Caresana, F.; Pelagalli, L. Hybrid system with micro gas turbine and PV (photovoltaic) plant: Guidelines for sizing and management strategies. *Energy* **2015**, *89*, 226–235. [CrossRef]
18. Arsalis, A.; Alexandrou, A.N.; Georghiou, G.E. Thermoeconomic modeling of a small-scale gas turbine-photovoltaic-electrolyzer combined-cooling-heating-and-power system for distributed energy applications. *J. Clean. Prod.* **2018**, *188*, 443–455. [CrossRef]
19. Darghouth, N.R.; Wiser, R.H.; Barbose, G.; Mills, A.D. Net metering and market feedback loops: Exploring the impact of retail rate design on distributed PV deployment. *Appl. Energy* **2016**, *162*, 713–722. [CrossRef]
20. Arsalis, A.; Alexandrou, A.N.; Georghiou, G.E. Thermoeconomic Modeling and Parametric Study of a Photovoltaic-Assisted 1 MWe Combined Cooling, Heating, and Power System. *Energies* **2016**, *9*, 663. [CrossRef]
21. Kélouwani, S.; Agbossou, K.; Chahine, R. Model for energy conversion in renewable energy system with hydrogen storage. *J. Power Sources* **2005**, *140*, 392–399. [CrossRef]
22. Darras, C.; Sailler, S.; Thibault, C.; Muselli, M.; Poggi, P.; Hoguet, J.C.; Melscoet, S.; Pinton, E.; Grehant, S.; Gailly, F.; et al. Sizing of photovoltaic system coupled with hydrogen/oxygen storage based on the ORIENTE model. *Int. J. Hydrogen Energy* **2010**, *35*, 3322–3332. [CrossRef]
23. Arsalis, A.; Nielsen, M.P.; Kær, S.K. Application of an improved operational strategy on a PBI fuel cell-based residential system for Danish single-family households. *Appl. Therm. Eng.* **2013**, *50*, 704–713. [CrossRef]
24. Mehr, A.S.; Gandiglio, M.; MosayebNezhad, M.; Lanzini, A.; Mahmoudi, S.M.S.; Yari, M.; Santarelli, M. Solar-assisted integrated biogas solid oxide fuel cell (SOFC) installation in wastewater treatment plant: Energy and economic analysis. *Appl. Energy* **2017**, *191*, 620–638. [CrossRef]
25. Kim, K.; Von Spakovsky, M.R.; Wang, M.; Nelson, D.J. A hybrid multi-level optimization approach for the dynamic synthesis/design and operation/control under uncertainty of a fuel cell system. *Energy* **2011**, *36*, 3933–3943. [CrossRef]
26. Kim, K.; Von Spakovsky, M.R.; Wang, M.; Nelson, D.J. Dynamic optimization under uncertainty of the synthesis/design and operation/control of a proton exchange membrane fuel cell system. *J. Power Sources* **2012**, *205*, 252–263. [CrossRef]
27. *Meteonorm 7*; Meteotest AG: Bern, Switzerland, 2012.
28. Open EI. Available online: https://openei.org (accessed on 7 May 2018).
29. Duffie, J.A.; Beckman, W.A. *Solar Engineering of Thermal Processes*, 4th ed.; Wiley: Hoboken, NJ, USA, 2013.
30. Arsalis, A. Thermoeconomic modeling and parametric study of hybrid SOFC-gas turbine-steam turbine power plants ranging from 1.5 to 10 MWe. *J. Power Sources* **2008**, *181*, 313–326. [CrossRef]
31. Kolb, G. *Fuel Processing for Fuel Cells*; Wiley-VCH: Weinheim, Germany, 2008.
32. Arsalis, A.; Nielsen, M.P.; Kær, S.K. Modeling and off-design performance of a 1 kWe HT-PEMFC (high temperature-proton exchange membrane fuel cell)-based residential micro-CHP (combined-heat-and-power) system for Danish single-family households. *Energy* **2011**, *36*, 993–1002. [CrossRef]
33. Arsalis, A.; Nielsen, M.P.; Kær, S.K. Modeling and optimization of a 1 kWe HT-PEMFC-based micro-CHP residential system. *Int. J. Hydrogen Energy* **2011**, *37*, 2470–2481. [CrossRef]
34. O'Hayre, R.; Colella, W.; Cha, S.-W.; Prinz, F.B. *Fuel Cell Fundamentals*, 2nd ed.; Wiley: Hoboken, NJ, USA, 2009.
35. Larminie, J.; Dicks, A.L. *Fuel Cell Systems Explained*, 2nd ed.; Wiley: Chichester, UK, 2003.
36. Arsalis, A.; Nielsen, M.P.; Kær, S.K. Modeling and parametric study of a 1 kWe HT-PEMFC-based residential micro-CHP system. *Int. J. Hydrogen Energy* **2011**, *36*, 5010–5020. [CrossRef]
37. Dincer, I.; Zamfirescu, C. *Sustainable Energy Systems and Applications*; Springer: New York, NY, USA, 2011.
38. GOV.UK Solar PV Cost Data. Available online: https://www.gov.uk/government/statistics/solar-pv-cost-data (accessed on 11 July 2018).

39. Scataglini, R.; Mayyas, A.; Wei, M.; Chan, S.H.; Lipman, T.; Gosselin, D.; D'Alessio, A.; Breunig, H.; Colella, W.G.; James, B.D. *A Total Cost of Ownership Model for Solid Oxide Fuel Cells in Combined Heat and Power and Power-Only Applications*; Lawrence Berkeley National Laboratory: Berkeley, CA, USA, 2015.
40. European Union Natural Gas Import Price. Available online: https://ycharts.com/indicators/europe_natural_gas_price (accessed on 3 July 2018).
41. *ASHRAE Handbook—HVAC Systems and Equipment*; American Society of Heating, Refrigerating, and Air-Conditioning Engineers: Atlanta, GA, USA, 2008.
42. Linderoth, S.; Larsen, P.H.; Mogensen, M.; Hendriksen, P.V.C.N.; Holm-Larsen, H. Solid Oxide Fuel Cell (SOFC) Development in Denmark. *Mater. Sci. Forum* **2007**, *539–543*, 1309–1314. [CrossRef]
43. Arsalis, A.; Alexandrou, A.N.; Georghiou, G.E. Thermoeconomic modeling of a completely autonomous, zero-emission photovoltaic system with hydrogen storage for residential applications. *Renew. Energy* **2018**, *126*, 354–369. [CrossRef]

© 2018 by the authors. Licensee MDPI, Basel, Switzerland. This article is an open access article distributed under the terms and conditions of the Creative Commons Attribution (CC BY) license (http://creativecommons.org/licenses/by/4.0/).

Article

Leveraging Energy Storage in a Solar-Tower and Combined Cycle Hybrid Power Plant

Kevin Ellingwood, Seyed Mostafa Safdarnejad, Khalid Rashid and Kody Powell *

Department of Chemical Engineering, University of Utah, Salt Lake City, UT 84112-9203, USA; k.ellingwood@utah.edu (K.E.); mostafa.safdarnejad@utah.edu (S.M.S.); khalidrashid85@gmail.com (K.R.)
* Correspondence: kody.powell@utah.edu; Tel.: +1-801-581-3957

Received: 29 November 2018; Accepted: 19 December 2018; Published: 24 December 2018

Abstract: A method is presented to enhance solar penetration of a hybrid solar-combined cycle power plant integrated with a packed-bed thermal energy storage system. The hybrid plant is modeled using Simulink and employs systems-level automation. Feedback control regulates net power, collector temperature, and turbine firing temperature. A base-case plant is presented, and plant design is systematically modified to improve solar energy utilization. A novel recycling configuration enables robust control of collector temperature and net power during times of high solar activity. Recycling allows for improved solar energy utilization and a yearly solar fraction over 30%, while maintaining power control. During significant solar activity, excessive collector temperature and power setpoint mismatch are still observed with the proposed recycling configuration. A storage bypass is integrated with recycling, to lower storage charging rate. This operation results in diverting only a fraction of air flow to storage, which lowers the storage charging rate and improves solar energy utilization. Recycling with a storage bypass can handle larger solar inputs and a solar fraction over 70% occurs when following a drastic peaking power load. The novel plant configuration is estimated to reduce levelized cost of the plant by over 4% compared to the base-case plant.

Keywords: concentrated solar power; hybridization; thermal energy storage; simulation; control

---

1. Introduction

Concentrated solar power (CSP) offers a potential path towards reducing carbon emissions from centralized power plants. Unlike photovoltaics, which convert solar energy directly to electricity, CSP utilizes the available solar thermal energy to drive conventional power cycles, such as the steam Rankine cycle [1]. CSP can be easily integrated with thermal energy storage (TES), which is advantageous compared to other energy systems that are limited to battery storage [2,3]. Because of these aspects, research and development of CSP has received a lot of attention in recent years.

Like battery storage with solar photovoltaics, TES allows for the solar energy to be stored for dispatch at a later time, generally when demand is higher [1,2,4]. TES is far less expensive than battery storage typically used to offset transient photovoltaics or other intermittent sources. Considerable work has been carried out to investigate TES integration into CSP plants [5–8] and there is widespread literature on the benefits and types of TES used in CSP plants [9–11]. Such work has been carried out with a goal of improving design and performance of TES systems such as enhancement of heat transfer when TES is charged and discharged in various configurations. Example works involve simulations of storage systems, exergy and economic analysis of their charging cycles, and validating these models with physical systems [12,13]. The configurations of CSP and TES systems can vary considerably. Currently, a two-tank storage system integrated into a parabolic trough collector (PTC) CSP plant is the most common configuration, both in physical and modeled systems [8]. However, novel configurations are still being developed even in PTC plants, and in recent years packed-bed storage has seen an increase in focus due to feasible integration into gas-driven Brayton cycles [14,15].

CSP can be easily hybridized with other energy sources, such as fossil fuels, due to equipment commonality. Through fossil fuel hybridization, the two energy sources operate in a synergistic way to produce power. This translates to more reliable operation of solar thermal plants, as well as increasing dispatchability of the converted energy to the grid during times of intermittent solar energy. Literature involving hybridization of CSP systems with natural gas is also quite abundant [6,16–18]. Hybridization offers an additional path forward to enhance solar energy utilization in CSP systems, as it has been shown to increase solar-to-electric efficiency (STE) relative to solar-only plants. STE is the marginal efficiency of converting available solar energy to electricity [19,20]. For a solar-only plant, plant operation may lower heat transfer and working fluid flow rates at times of lower solar activity so that the design temperatures can be met. Lower flow rates result in off-design operation and can lead to stagnation and dips in overall plant efficiency. For air turbine plants where air acts as both the heat transfer and working fluid, hybridization sustains turbine firing temperatures without having to lower the flow rate of the heat transfer fluid during hours of lower sunlight [19,21]. The sustained flow rates mitigate thermal losses in the solar collector and result in higher STE versus plants operating via only solar energy. Additionally, hybridization continuously maintains design temperatures during hours of intermittent solar activity or during delays caused by storage discharge. This steady temperature control further contributes to overall system performance and solar energy utilization [22]. A solar-tower, hybrid combined cycle plant has been proposed in literature [23–25] and a more common solar-tower Brayton plant has been developed at the pilot level [26–28] and modeled extensively [5,18,29]. Studies involving tower-driven plants have seen an increase in interest due to their high concentration ratios, which can realize the high operating temperatures needed to drive the Brayton cycle and combined cycle plants [14]. Both systems require temperatures greater than what PTC plants can realize, unless the integrated solar combined cycle (ISCC) is considered. Within ISCC systems, CSP compliments operation of the bottoming Rankine cycle, which can be realized using a PTC system [30,31].

While physical improvement of TES and hybrid design is vital for development, operation and automation of CSP plants represent substantial efforts for studies involving TES integration and hybridization. Advanced process control (APC) has been applied to common CSP equipment to improve operation [21,32,33]. In all CSP systems, the collector exit temperature is a primary control variable [34]. Camacho et al. have worked extensively to apply several APC methods, such as model predictive control, to regulate exit temperatures in trough collectors [35]. Similarly, cascade control has also been used to control the exit temperature of such collectors [36]. In recent years, many APC has been applied to automate heliostat field operation for central tower receivers, with such studies focused on controlling solar flux distribution at the receiver and maximizing energy collected by the tower receiver [34]. These control schemes typically involve manipulation of heliostat orientation to maintain solar flux on the receiver to regulate the collector exit temperature. If the temperature becomes too high, mirror orientations are modified to direct irradiance away from the receiver. In tower systems, these control schemes monitor the position of the sun and shifting of the heliostats to distribute high solar fluxes to the collector surface, where a heat transfer fluid absorbs the solar irradiance [37].

While significant work has focused on component-level operation of CSP plants, such as control of collection temperature, there remains much opportunity to improve CSP performance by focusing on systems-level control and operation. Such holistic methods can help to improve solar penetration by employing sophisticated control algorithms to leverage TES integration and hybrid operation [38]. Of interest in this work is to employ holistic automation to realize a high solar fraction in a hybrid power plant. The inclusion of the aforementioned storage is not enough to realize higher solar fractions. Smart, holistic automation must be utilized to achieve this goal [6,21,39]. This is done by designing plant configurations and developing control algorithms that can direct excess solar energy to heat sinks, such as the packed-bed TES considered in this study, to better harness excess solar energy without having to direct it away from the collector. In the system presented herein, a recycling configuration is proposed as an alternative means to control receiver exit temperature by increasing thermal capacity of the heat transfer fluid during periods with high solar activity. The recycle stream can redirect excess energy to the packed bed storage system to be dispatched at a later time. This study would be the first to develop such an operation of a hybrid solar-combined cycle power plant with energy storage with a goal to increase solar fraction under a peaking power production schedule. This is achieved by systematically changing the plant configuration as well as the control schemes and analyzing the advantages and disadvantages of each configuration. This work demonstrates that by employing systems-level automation, the solar fraction can be significantly boosted while maintaining collection temperature control. The proposed control schemes also demonstrate tight power control, while diversion of solar energy away from the heliostats, as proposed by previous research, is avoided altogether.

From here on the paper is organized as follows. Section 2 describes the various plant configurations developed and studied: a base-case configuration, a configuration utilizing a recycle stream to control receiver temperature, a configuration utilizing the recycle stream and storage bypass to enhance stored energy dispatch. Examples of system dynamics are also presented here with context to the control scheme of the recycling configuration. The model equations for these configurations are then presented in Section 3. Section 4 discusses the dynamics of each scheme with regards to robust control and solar energy utilization. This section highlights a study to maximize the solar fraction of the plant and test the plant's ability to handle large amounts of solar input. Additionally, an economic analysis of the final plant configuration is presented in comparison to the base-case power plant. Finally, Section 5 presents the conclusion of the study.

2. Overview of Plant Configurations

The hybrid solar-combined cycle plant has a capacity of roughly 200 MW between Brayton and Rankine cycles. Three configurations for plant operation are presented here (Figures 1–3). The control scheme shown in Figure 1 is used as a basis for the plant. Ambient air is compressed and passed along to a central tower receiver (CTR) where a heliostat field directs concentrated sunlight to heat flowing air during hours of sunlight. When there is no solar activity, the air bypasses the receiver. Downstream of the collector, air passes through a packed-bed storage system. Then, the air passes through a combustion chamber and the hot flue gas is then directed to a gas turbine to produce power. The turbine firing temperature is controlled continuously by manipulating the natural gas flow into the combustion chamber. The net power of the plant is controlled by manipulating the inlet guide vane (IGV) angle of the compressor, which dictates air flow through the system on a volumetric basis. The IGV angle also controls the exit temperature of the CTR. Because the IGV controls both the receiver exit temperature and net power, a high-value selector (HVS) control is employed. The HVS continuously selects the highest IGV angle input between the power and CTR temperature control loops. The primary objective of the HVS controller is to maintain the power setpoint but it can override that setpoint, so collector temperature does not exceed design values.

The operation of the base-case plant allows for some flexibility in plant operation, but at some point, the amount of solar energy input exceeds the thermal capacity of the air heat transfer fluid. This results in off-design temperatures within the receiver even with the HVS override. To mitigate this and to harness the excess solar energy utilizing the physical sinks in the system, a recycling configuration is proposed, as shown in Figure 2. Recycling increases the air flow and thus the thermal capacity of the heat transfer fluid in the receiver. This configuration presents a possible means to control CTR exit temperature as an alternative to redirecting excess solar energy to ambient heat (essentially losing solar energy). This solution to temperature control developed in this study is similar to previous research focusing on temperature control in parabolic trough systems through fluid flow manipulation [40]. Figure 4 shows the dynamics of the HVS-recycle control scheme for the recycling configuration. Initially, as direct normal irradiance (DNI) increases throughout the day (Figure 4A), the recycle stream (Figure 4B) turns on to control the receiver exit temperatures (Figure 4D) by increasing mass flow through the tower. Once the recycling temperature, or storage exit temperature, reaches 1000 K (Figure 4E), the recycle stream turns off. When the recycle loop turns off, the temperature control of the HVS may override the power control (Figure 4C) if the solar energy still available would result in temperatures higher than design conditions. The temperature setpoint of the recycle and HVS control loops, shown in Figure 4D, must be offset so that they do not interfere with one another; that is, the setpoint of the temperature from recycling is lower than the HVS control so that the HVS does not control collector temperature while recycling occurs.

With the increased thermal capacity of the flowing air, an increase in field size or an increase in solar activity results in an increased solar fraction while also maintaining power and temperature of the system. Without a recycle stream, the base-case configuration is not capable of maintaining both power and collector temperature. However, at some point of incrementally increasing field size, the introduced solar energy begins to exceed the thermal capacity of airflow even with the additional recycling capacity. Thus, excessive temperatures are realized in the collector. While there is still substantial solar activity left in the day, the recycle stream is no longer active, as the temperature of the recycle stream has reached its limit of 1000 K. Additionally, as the storage charges, recycle stream steadily increases in temperature at the TES exit, which in turn elevates the temperature entering the tower, making it harder for additional flow to maintain the temperature setpoint. To mitigate the rate at which the recycling temperature rises, a bypass of the TES is implemented in conjunction with recycling. This plant configuration is seen in Figure 3. The bypass operates using on/off control logic tied to the recycle stream:

If the recycle loop is active to keep receiver temperature at 1300 K, the bypass stream is active. The air flow due to IGV angle bypasses the storage and the recycle stream charges the storage.

If the recycle loop is inactive, the bypass stream is inactive as well. The intake air (which represents the total air flow when not recycling) passes through the TES to discharge any stored energy. Otherwise, if there is no stored energy, the storage is bypassed. The recycle loop turns off in one of two ways:

a. The temperature of the recycle loop (storage exit temperature) reaches 1000 K. This prevents already heated air from entering the receiver. If this is the case and there is still solar activity which would result in excess temperatures, the HVS operates in temperature control.
b. The receiver operates below the 1300 K setpoint and temperature control is not needed.

The bypass allows for only a fraction of the air flow to charge the TES while the recycle stream is active, which results in a lower TES charging rate and rise of TES exit temperature. This results in a longer period that the recycle stream can be active to control temperature and absorb excess solar energy. Without the bypass, the flow from both recycle and inlet charge the TES and the elevated exit temperature is realized far more quickly.

In all plant configurations, the operation of the bottoming steam cycle is identical: the steam cycle operates by controlling the temperature of the steam leaving the heat recovery steam generator (HRSG). The power from the Rankine cycle is not directly controlled but is estimated from the flow needed to maintain a temperature setpoint of steam at the HRSG outlet. Prior to the steam cycle, the flue gas undergoes auxiliary firing to reach an operating temperature of 820 K [41].

To observe the benefits of TES within the system and to more realistically represent variable grid demand where the load is higher during evenings, a peaking power operation is proposed and used as a basis for this study. Under this schedule, the power is set at 100 MW from 11 p.m. to 11 a.m. At 11 a.m., the on-peak power setpoint is set at 175 MW for the next 12 h. The 175 MW is not quite at capacity so that there is room for the IGV angle to increase and control temperature control if needed.

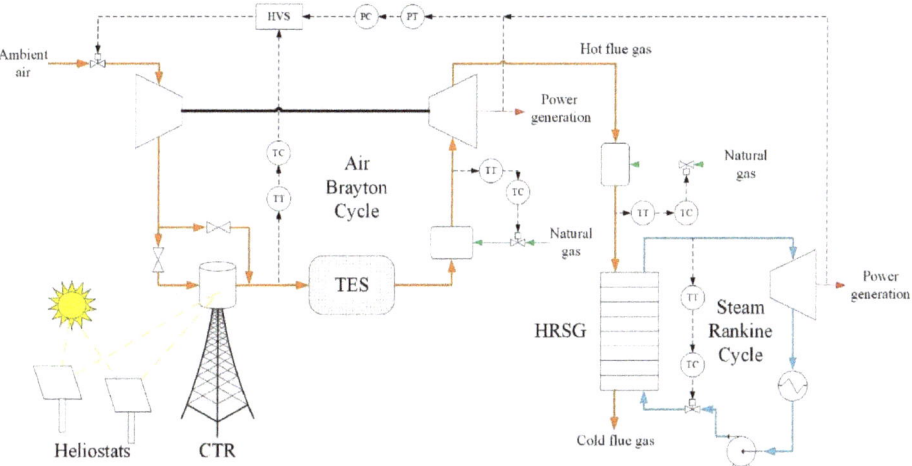

Figure 1. Schematic of the base-case configuration of the power plant.

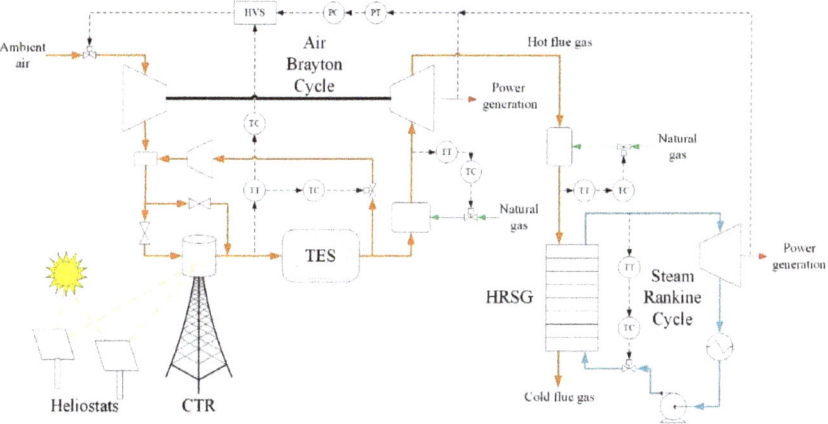

Figure 2. Schematic of the recycling configuration of the power plant.

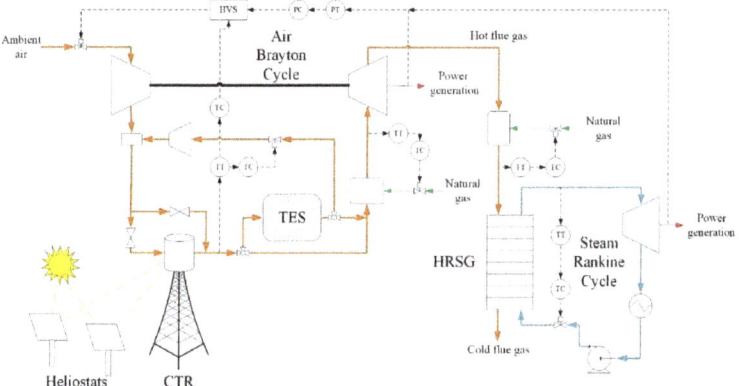

Figure 3. Schematic of the recycling with thermal energy storage (TES) bypass configuration of the power plant.

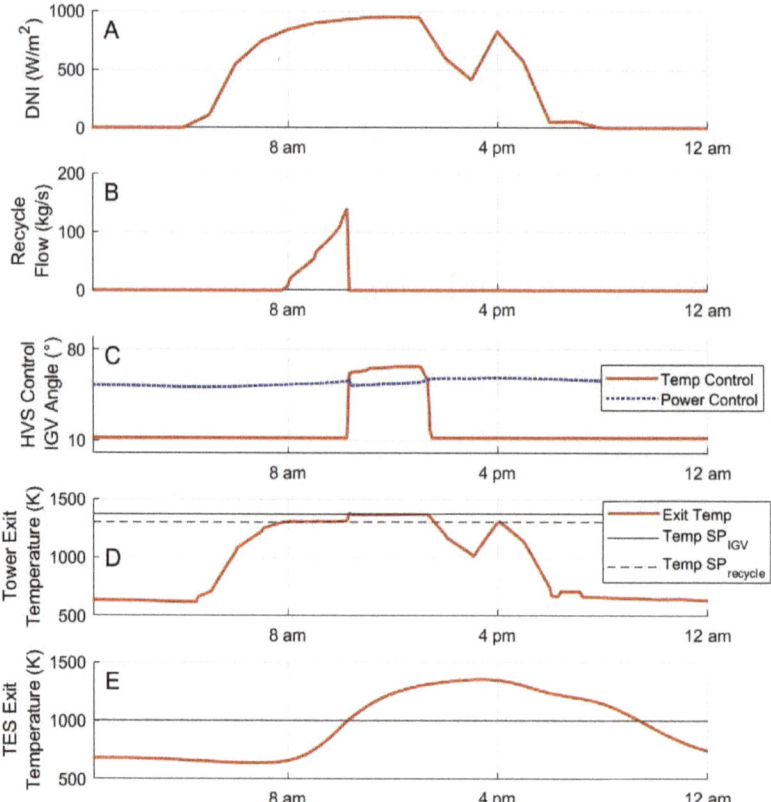

Figure 4. Dynamics of high-value selector (HVS)-recycle control for a plant set at 150 MW load and utilizing only a recycle control (no storage bypass). From top to bottom: (**A**) direct normal irradiance (DNI) for June 24th; (**B**) recycle loop flow rate; (**C**) inlet guide vane (IGV) angle dictated by power/temperature HVS controller; (**D**) exit temperature of central tower receiver (CTR), (**E**) storage exit temperature along with 1000 K limit for recycling temperature. SP = setpoint.

For the recycling configuration, a block diagram of the power plant is seen in Figure 5. This diagram shows the feedback control loops implemented in the recycling plant and is meant to show the flow of information within the control algorithm. The diagram omits the receiver and TES bypass streams for simplicity purposes.

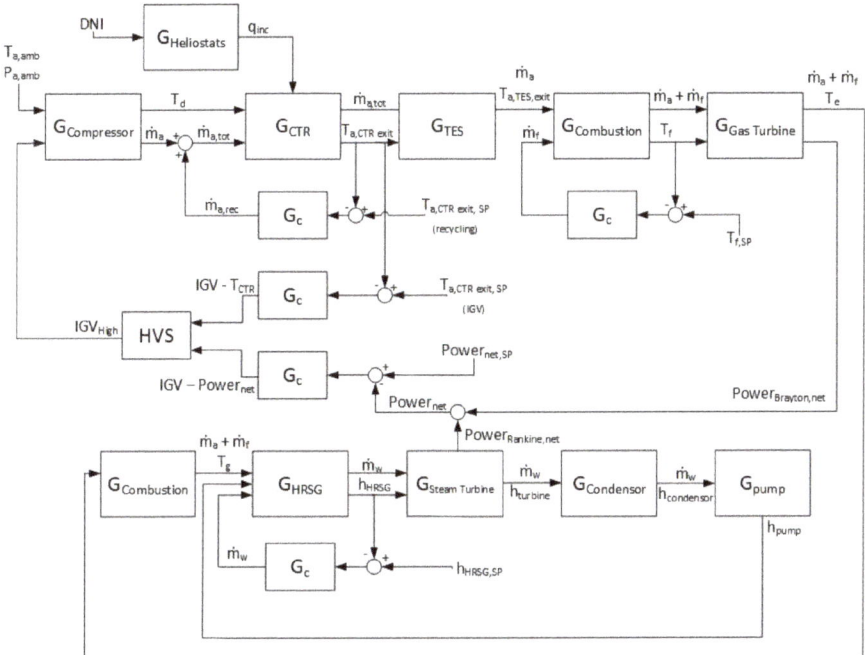

Figure 5. Block diagram of the power plant employing a recycling configuration. The variables of the diagram are described in Nomenclature.

3. Modeling and Methodology

The hybrid plant model is developed in Simulink using first principles [42–45]. This section presents a general overview of the plant model.

3.1. Gas Turbine Components

The models of the Brayton cycle components are assumed to be relatively fast compared to the transient solar components and steady-state, thermodynamic models are used for Brayton cycle components. These gas turbine components are:

3.1.1. Compressor

The compressor model simulates the inlet air flow rate and compressor exit temperature:

$$T_d = T_{amb} \times \left(1 + \frac{x_c - 1}{\eta_c}\right) \quad (1)$$

$$x_c = \left(PR_C \times \frac{\dot{m}_a}{\dot{m}_{a,n}}\right)^{\frac{\gamma_c - 1}{\gamma_c}} \quad (2)$$

$$\dot{m}_a = \frac{P_{amb}}{P_{amb,o}} \sqrt{\frac{T_{amb,o}}{T_{amb}}} \frac{\sin(\theta_{IGV} - \theta_{min})}{\sin(\theta_{max} - \theta_{min})} \quad (3)$$

where $\dot{m}_a$ is the mass flow rate of air dictated by the IGV angle (θ_{IGV}), T_d is the compressor exit temperature, T_{amb} is the compressor inlet temperature or ambient temperature, η_c is the compressor efficiency, PR_C is the compression ratio, $\dot{m}_{a,n}$ is the nominal intake air flow rate, and γ_c is the cold end ratio specific heats. P_{amb} is the atmospheric pressure, $P_{amb,o}$ is the reference ambient pressure,

$T_{amb,o}$ is the reference ambient temperature, θ_{min} is the minimum IGV angle, and θ_{max} is the maximum IGV angle.

3.1.2. Combustion Chamber

The exit temperature of the combustion chamber or the turbine firing temperature, T_f, is calculated from the following equation:

$$T_f = T_d + \left(\frac{\eta_{comb} LHV}{c_{p,h}}\right)\left(\frac{\dot{m}_f}{\dot{m}_f + \dot{m}_a}\right) \tag{4}$$

where η_{comb} is the efficiency of combustion, LHV is the lower heating value of natural gas, $c_{p,h}$ is the specific heat of the exhaust gas flow, and $\dot{m}_f$ is the fuel (natural gas) flow rate.

$$T_e = T_f\left(1 - \left(1 - \frac{1}{x_h}\right)\eta_t\right) \tag{5}$$

$$x_h = \left(PR_T \times \frac{\dot{m}_f + \dot{m}_a}{\dot{m}_{f,n} + \dot{m}_{a,n}}\right)^{\frac{\gamma_h - 1}{\gamma_h}} \tag{6}$$

where PR_T is the compression ratio of the turbine, η_t is the turbine efficiency, $\dot{m}_{f,n}$ is the nominal fuel flow, and γ_h is the hot end ratio of the specific heats. From the gas turbine, the hot flue gas is used to power a steam cycle [44].

3.2. Central Tower Receiver

Upstream from the natural gas firing, the CTR heats the compressed air. The tubular gas receiver model consists of energy balances to simulate temperature profiles of flowing air, carrier pipes (tubes), and glass casing, which vacuum-seals the fluid pipes from atmospheric conditions. Dynamic energy models are presented for the three components using control volume energy balances, considering convective and radiative heat transfer. The models assume uniform radial temperature profiles and neglect conductive heat transfer. The energy models are discretized to simulate the axial temperature profile along the length of the tower collector system using integrator function blocks within Simulink. The energy models proposed are:

Air (a):

$$\rho V_a c_{p,a} \frac{\partial T_a}{\partial t} = \dot{m}_a c_{p,a} \Delta T_a - h_i A_p (T_a - T_p) \tag{7}$$

Pipes (p):

$$\rho_p V_p c_{p,p} \frac{\partial T_p}{\partial t} = h_i A_p (T_a - T_p) - \frac{\sigma\left(T_p^4 - T_c^4\right)}{\frac{1-\varepsilon_p}{\varepsilon_p A_p} + \frac{1}{A_p} + \frac{1-\varepsilon_c}{\varepsilon_c A_c}} + q_{inc} A_c \tau_c \nu_p \tag{8}$$

Glass case (c):

$$\rho_c V_c c_{p,c} \frac{\partial T_c}{\partial t} = \frac{\sigma\left(T_p^4 - T_c^4\right)}{\frac{1-\varepsilon_p}{\varepsilon_p A_p} + \frac{1}{A_p} + \frac{1-\varepsilon_c}{\varepsilon_c A_c}} - h_o A_c (T_c - T_{amb}) - \varepsilon_c \sigma A_c \left(T_c^4 - T_{amb}^4\right) \tag{9}$$

where ΔT_a represents the change in temperature of flowing air through the discretized shell volume. The terms ρ_j, $c_{p,j}$, h_i, and h_o are the density, heat capacity, internal flow transfer coefficient between the air and piping, and external heat transfer coefficient between ambient air and the glass casing, respectively. Temperature dependency of these terms is considered. The heat transfer coefficients are estimated from correlations given by [46] under internal turbulent flow regimes and flow over a flat plate (external flow across the receiver panels). Additionally, V_j, T_j, A_j, t, x represent the shell volume, shell temperature, heat transfer area, time, and axial position in the tube, respectively, while emissivity, transmissivity, and absorptivity for component j are represented by ε_j, τ_j, and ν_j. The Stefan-Boltzmann constant is also shown by σ and q_{inc} is the solar irradiance incident on the receiver, which is simulated using typical meteorological yearly data and a heliostat field model developed in previous work [47].

3.3. Thermal Energy Storage

A thermal energy storage model simulates the temperature of the working fluid air and stone medium:

Air (a):
$$\rho_a V_a c_{p,a} \frac{\partial T_a}{\partial t} = \dot{m}_a c_{p,a} \Delta T_a - h_s A_s (T_a - T_s) \tag{10}$$

Stone (s):
$$\rho_s V_s c_{p,s} \frac{\partial T_s}{\partial t} = h_s A_s (T_a - T_s) \tag{11}$$

where A_s is the surface area of stone and h_s is the temperature-dependent heat transfer coefficient between the packing and air. Like the CTR model, the TES model is discretized along the axial direction. A correlation for the heat transfer coefficient in a packed bed is used from [2,41,48].

3.4. Steam Cycle

The HRSG is modeled as a heat exchanger using the Number of Heat Transfer Units (NTU)-effectiveness method [46,49] The NTU-effectiveness method approximates heat transfer based upon the maximum possible heat transfer rate. The maximum heat transfer rate is defined as:

$$Q_{max} = C_{min}(T_{hot,in} - T_{cold,in}) \tag{12}$$

where C_{min} is the minimum heat capacity rate of the two fluids involved, which are in this case, the hot flue gas and water/steam. $T_{hot,in}$ is the hot flue gas inlet temperature, $T_{cold,in}$ is the inlet temperature of saturated water. The minimum heat capacity rate limits the maximum amount of transferable energy between the participating mediums. The heat capacity rate of a species is its mass flow rate multiplied by its heat capacity. Because one of the mediums undergoes a phase change (steam), the heat capacity rate of the single-phase medium (flue gas) limits the heat transfer and so $C_{min} = (\dot{m}_f + \dot{m}_a)c_{p,h}$. The energy transferred in the HRSG is found and used to determine the change in temperature of the flue gas and the production rate of steam assuming a constant evaporation enthalpy:

$$\begin{aligned} Q_{HRSG} &= \epsilon Q_{max} \\ Q_{HRSG} &= \dot{m}_w \Delta h_w = (\dot{m}_f + \dot{m}_a) c_{p,h} \Delta T_g \end{aligned} \tag{13}$$

where ϵ is the effectiveness, $\dot{m}_w$ is the water flow rate, Δh_w is the enthalpy change of water as it converts to steam, ΔT_g is the change in temperature of the flue gas. The effectiveness of the heat exchanger is assumed to be 0.8. The HRSG is assumed to be isobaric and operating at 75 bar on the water side. Using interpolation of steam table data [50], the outlet enthalpy, temperature, and entropy of steam are found at outlet conditions based upon Equation (13).

The turbine and pump were both assumed to be isentropic, and outlet conditions are estimated at the inlet entropy values. Similarly, the outlet conditions of the condenser are found assuming ideal phase change over a constant pressure. The work of the pump and turbine, along with the rejected heat of the condenser are all found using enthalpy energy models similar to the one used in the HRSG model:

$$W_{turbine} = \dot{m}_w(h_{HRSG,exit} - h_{turbine,exit}) \quad (14)$$

$$Q_{condenser} = \dot{m}_w(h_{turbine,exit} - h_{condensor,exit}) \quad (15)$$

$$W_{pump} = \dot{m}_w(h_{condenser,exit} - h_{pump,exit}) \quad (16)$$

The exit enthalpies for each component were approximated based upon the process conditions and simulated using steam table data. The turbine and pump are both assumed isentropic and the outlet conditions, now at different pressures from the inlet, correspond to the inlet entropy conditions. The condenser, like the HRSG, is isobaric and exit enthalpy is found assuming constant pressure from inlet conditions.

3.5. System Power

The net power generated is the combined net power from the air-gas turbine and steam turbine where compression work is accounted for:

$$Power_{net} = P_{Brayton,net} + P_{Rankine,net} \quad (17)$$

Depending on the system configuration, the net power will change. For the base-case plant seen in Figure 1, the net power is:

$$Power_{net} = \left[(\dot{m}_f + \dot{m}_a)c_{p,h}(T_f - T_e) - \dot{m}_a c_{p,a}(T_d - T_{amb})\right] + [\dot{m}_w \Delta h_{turbine} - \dot{m}_w \Delta h_{pump}] \quad (18)$$

For the systems utilizing a recycle stream, the recycle compressor needs to be accounted for in net power production:

$$Power_{net} = \left[(\dot{m}_f + \dot{m}_a)c_{p,h}(T_f - T_e) - \dot{m}_a c_{p,a}(T_d - T_{amb}) - \dot{m}_{a,rec}c_{p,a}(T_{d,rec} - T_{TES,exit})\right] \\ + [\dot{m}_w \Delta h_{turbine} - \dot{m}_w \Delta h_{pump}] \quad (19)$$

where $c_{p,h}$ is the heat capacity of the turbine exhaust gas [43].

3.6. System Performance Parameters

To evaluate the performance of each configuration, certain metrics are defined. The overall plant efficiency is defined as the net energy produced relative to the available energy from both solar and natural gas. The available solar energy is based upon the field collector area (A_{field}) and direct normal irradiance. The available natural gas energy is based upon the mass flow and the LHV of natural gas. The overall efficiency is:

$$\eta_{overall} = \frac{E_{total}}{E_{solar,available} + E_{gas,available}} = \int \frac{Power_{net}}{A_{field}DNI + \dot{m}_f LHV} dt \quad (20)$$

The solar performance is evaluated using the solar fraction (SF) and STE. The solar fraction is the amount of generated energy that comes directly from solar energy utilization:

$$SF = \frac{E_{solar,utilized}}{E_{total}} = \int \frac{Power_{net} - \eta_f \dot{m}_f LHV}{Power_{net}} dt \quad (21)$$

STE is the marginal energy production from solar relative to the total available solar energy and is calculated using:

$$STE = \frac{E_{solar,utilized}}{E_{solar,available}} = \int \frac{Power_{net} - \eta_f \dot{m}_f LHV}{A_{field} DNI} dt \qquad (22)$$

4. Results and Discussions

The results are presented as follows. First, the performance of the recycle configuration is compared to that of the base-case configuration. The advantages of recycling are shown by highlighting robust control and improved solar energy utilization. Second, the introduction of a TES bypass with recycling is shown to further improve solar energy utilization without losing system control. Third, the scheme with bypass and recycle streams is further tested using a peaking power load. Lastly, an economic analysis is presented to show the improved levelized cost due to the final plant configuration. For all plots showing system dynamics, the day of June 24th was decided upon as it represents a day with substantial and dynamic solar activity for the location of Las Vegas [51].

4.1. Recycle vs. Base-Sase

A major goal for this work is to develop an operation scheme with the associated control loops to maximize the solar fraction of the plant. This is achieved by employing systems-level control to reliably dispatch energy through hybridization and storage integration. As mentioned, previous work in the literature has focused on mirror orientation to maintain collector temperatures but through manipulation of air flow and leveraging the packed-bed TES, excess solar energy can be harnessed without having to direct solar energy away from the collector. The obvious means to increase solar fraction is to increase the heliostat field size and storage in accordance with the field. However, under the base-case configuration seen in Figure 1, during days of high solar activity, excess temperatures are sometimes realized even with the HVS override. In other words, the thermal capacity of the heat transfer fluid has been saturated relative to the amount of solar energy available, and the operating temperature exceeds the design specification. Figure 6 compares the dynamics of the base-case and the recycling configurations with respect to net power and temperature control. The plot also shows the manipulated variables for each control scheme and solar activity as a reference. As observed from Figure 6, the base-case configuration is able to maintain the temperature of the air leaving the tower using the HVS controller. However, because the IGV angle must increase to mitigate excessive temperatures during times of high solar activity, the power control is overridden, and the net power output increases accordingly, not allowing the plant to maintain power control.

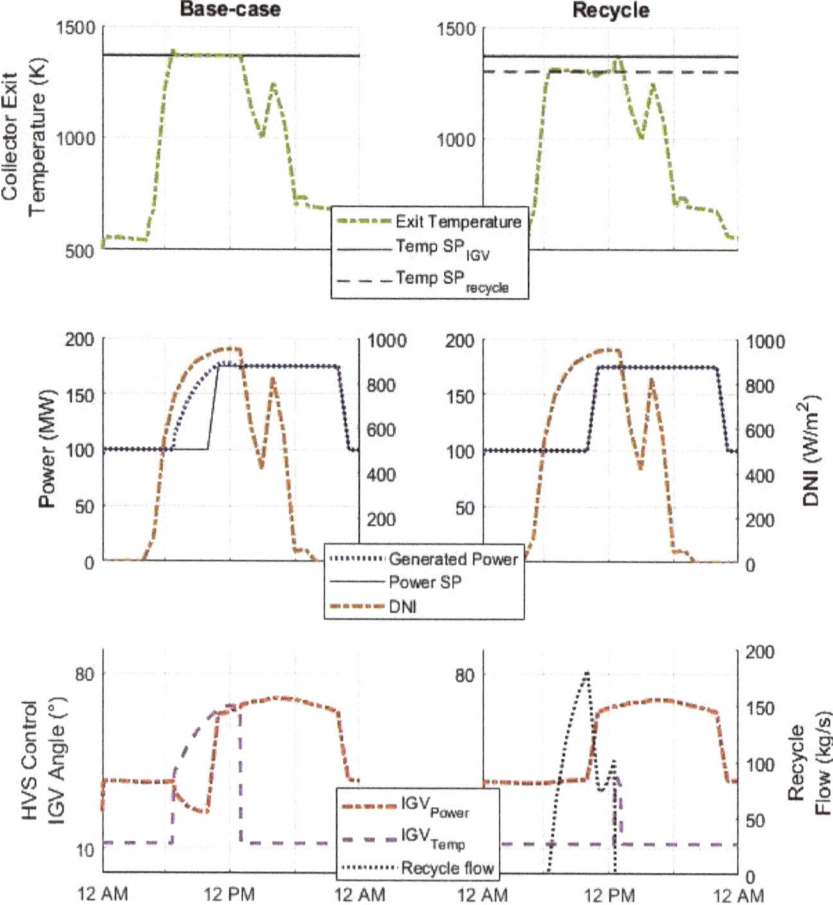

Figure 6. Temperature and power control for the base-case and recycling plants for June 24th (SP = setpoint).

Air recycling enables independent control of the collector exit temperature and net power production, so long as recycling is viable. Recycling is an option so long as the TES exit temperature is lower than designated recycling threshold. This decoupling of control mitigates excessive temperatures within the tower collector without losing robust power control, as can also be seen in Figure 6. Without recycling, the HVS control would otherwise have to override power control. This improved control enhances solar energy utilization and offers flexibility to increase collector field size to improve solar fraction.

Table 1 shows the yearly performance comparison of the base-case to the recycling configuration. Both plants follow the proposed peaking schedule, utilize a heliostat field size at 550,000 m² and contain a TES sized at 5700 m³ or roughly 6 h [31]. Use of the recycle results in a slight decrease in overall plant efficiency and STE when compared to the base-case plant which is likely due to an increase in receiver inlet temperature during active solar hours. The increase in inlet temperature results in larger fractional thermal losses due to radiation. However, the recycle-loop plant does perform better in terms of the solar fraction. Because the base-case plant must further open IGV's to maintain receiver temperature, the total amount of energy (from both solar and gas) produced is larger than the total energy produced from the recycle-controlled plant, despite the same input of solar energy. Therefore, it is logical that the recycling configuration exhibits a larger solar fraction compared to the base-case as the solar input is identical for both plants. This finding further highlights the ability to continuous control plant power when recycling without having to go off setpoint to maintain collector temperatures.

Table 1. Yearly performance comparison for the base plant vs. plant with recycle stream.

Metric	Base-Case (%)	Recycle (%)	Change Relative to Base-Case (%)
η	37.8	36.3	−3.9%
STE	22.4	22.2	−0.8%
SF	29.9	31.4	+4.9%

4.2. Recycle + Bypass vs. Recycle Only

Even with a recycle stream, attempting to increase solar fraction by increasing field size has limitations in plant performance similar to the base-case plant. As the storage charges, the increasing storage exit temperature has a heightened effect on the inlet temperature of the receiver as the recycling temperature increases the temperature entering the receiver. The rising exit temperature requires more flow to be recycled to keep receiver temperatures at design specifications. Therefore, a configuration utilizing a storage bypass in conjunction with the recycle loop (Figure 3) is proposed. Under this configuration, only the recycle flow passes through the TES while recycling is active. This operation means that the storage is charged at a significantly lower rate. The plant utilizing only a recycle loop (Figure 2) charges the storage using both the recycle and the air flow rate from the IGVs. This lower charging rate when bypassing TES translates to mitigation of excessively high recycle temperatures, which allows the recycle stream to be active for a longer time when temperature control is needed. This translates to even greater flexibility to harness excess solar energy while maintaining power control.

The dynamics of the plant utilizing a fractional bypass are shown in Figure 7 compared to the dynamics of the plant with only a recycle stream. The plots show the dynamics for both plants employing a field sized at 680,000 m^2, which is roughly a 24% increase in size from the aforementioned comparison of the base-case vs. recycling configurations. The increase in field size is used in lieu of cases where a significant amount of solar energy is available.

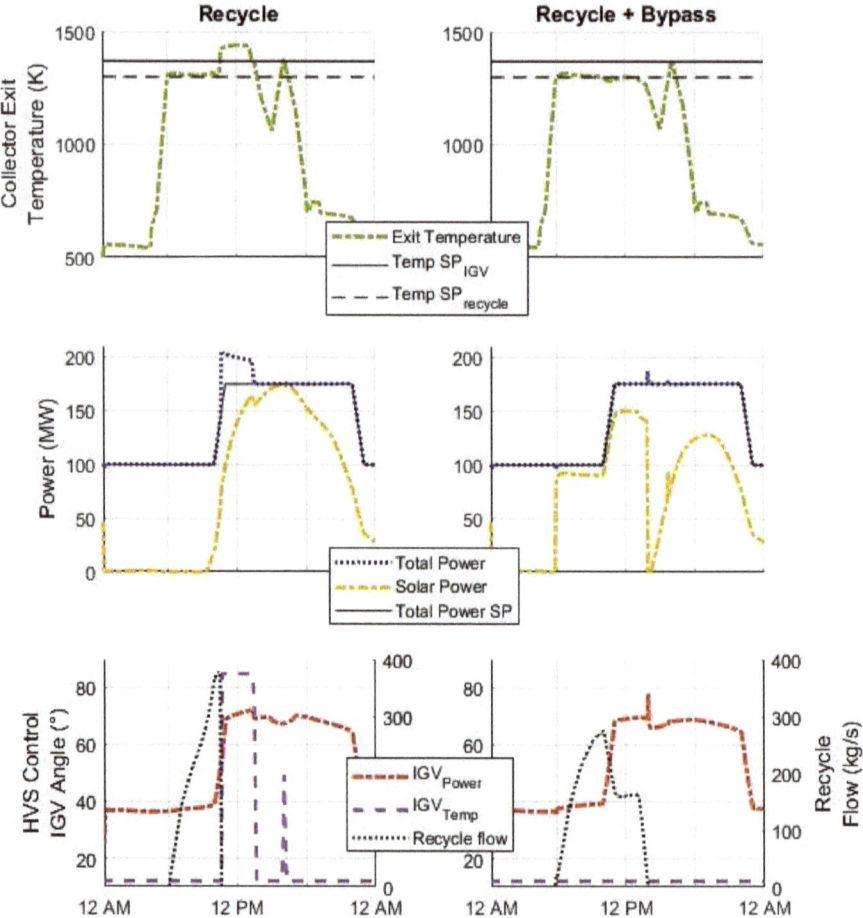

Figure 7. Comparison of plant control with and without TES bypass for June 24th.

Similar to the base-case plant, the recycle-only configuration struggles to maintain the power setpoint when a larger field is deployed. The storage charges quickly and the HVS selector control overrides the power control of the system in favor of temperature control. However, even at the maximum IGV angle, the thermal capacity of the working fluid is too small for solar energy being introduced to the system. For the bypass system, the slower rate of charging is evident by the longer period that the recycling stream is active. Without bypass, the temperature of the recycle stream rises quickly and forces the recycle stream to shut off under the recycle-only operation. With the bypass, the power setpoint is maintained throughout the day as the HVS control is not needed to control the temperature. Once the recycle/bypass control is turned off, the total air flow (inlet) of the plant passes through the storage, charging the storage if there is still a positive difference between the receiver and storage temperature, and finally discharging the storage once inlet temperatures are lower than storage temperatures.

As can be seen in Figure 7, the solar power produced occurs in two distinct phases when the storage bypass is employed. The first phase follows a pattern similar to the overall power production. During this time, a fraction of the absorbed solar energy, which is due to recycling flow, is dispatched to storage for later use. The remainder (due to IGV) is dispatched directly to the combustion chamber and turbine, which results in a similar pattern as the overall power production. The storage discharging cycle can be examined during the second phase of solar power production, starting around 2 pm. Prior to the second phase, a dip in solar production occurs, which is a result of the temperature gradient in the storage and the time needed for the elevated temperatures at the beginning of the storage to propagate through the bed via advection. Ideally, a hybrid plant would need to be developed so that it can eventually operate primarily through solar usage and continually maintain power without seeing dips in solar production as seen here. The dip in temperature additionally showcases the importance of having hybridized fuel usage so that design temperatures can be maintained while delayed dispatch from the storage is limiting the temperature profile downstream from TES. It exemplifies the reliability seen in a hybrid plant that can dispatch energy quickly when necessary. Furthermore, having to deal with energy propagation through the packed-bed storage poses a potential issue for a plant to operate with consistent solar energy production. This phenomenon gives an opportunity to investigate possible apply advanced control methods to overcome this advective delay that can cause the dip in solar energy generation. Otherwise, a switching system is needed where the directional flow of charging and discharging is opposite, which has been presented by [6]. However, having a switching method would need additional equipment resulting in increased capital cost. Approaching this dip from a process control perspective may allow for consistent operation while mitigating some of the cost of equipment and storage design.

Table 2 shows the yearly performance comparison of the plant operating via recycle and bypass versus operation under recycle-only. Interestingly, the plant using a bypass operates at a slightly lower overall efficiency, while maintaining STE. This is explained by the increased solar fraction that the bypass system exhibits. Within hybrid CSP systems, there exists a tradeoff between solar fraction and overall efficiency. As solar fraction increases, the overall efficiency decreases [52], and findings here further exemplify that tradeoff. However, despite a dip in overall efficiency, the STE of the configuration with both recycle and bypass is similar to the recycle-only plant. Overall, the addition of storage bypass allows for substantial improvements in solar harnessing with respect to the solar fraction. While the use of only a recycle stream to control receiver temperature is valid, utilization of a bypass stream represents an additional means for flexible plant operation while maintaining power load.

Table 2. Yearly performance comparison for recycle-only plant vs. recycle plant with bypass.

Metric	Recycle-Only (%)	Recycle + Bypass (%)	Change Relative to Recycle Only (%)
η	36.5	35.9	−1.8
STE	23.6	23.6	+0.0
SF	36.6	39.1	+6.6

4.3. Solar Fraction Enhancement

More drastic peaking schedules are also tested to potentially observe very high solar fractions. These scenarios are carried out via a parametric study at incrementally lower power sets point during non-peak hours following the peaking schedule, as previously mentioned. Initially, simulations ran with a field sized at 680,000 m² and a 6-h TES. This size of field and storage corresponds to the same size of the components studied in Section 4.2. The off-peak power setpoint is decreased incrementally with values beginning at 100 MW and simulations are run at each of those power schedules for the day of June 24th. Lowering of the off-peak setpoint is continued until the plant was unable to maintain the power load due to the HVS-control scheme favoring temperature control over power. The plant performances are summarized in Figure 8.

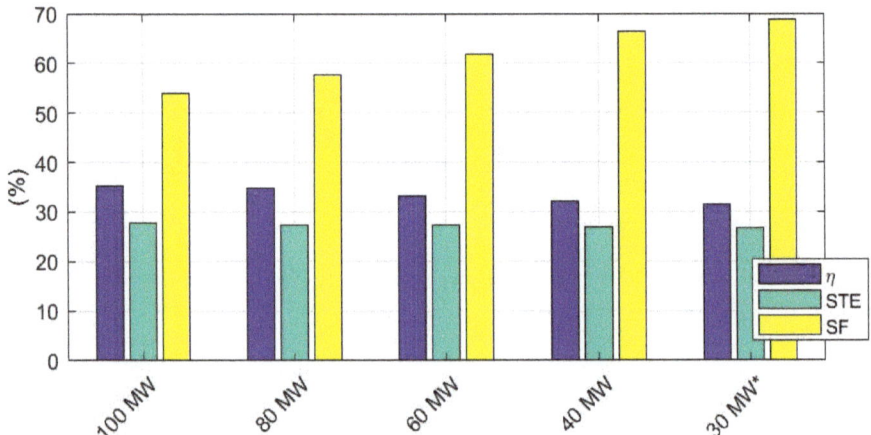

Figure 8. Summary of the parametric study of sequential decreases in off-peak power setpoint for peaking scenario. * The plant could not maintain power load at a 30 MW lower setpoint.

As seen in Figure 8, lowering the power setpoint during non-peak hours can significantly improve the solar fraction of the plant. Here, the plant exhibits a daily solar fraction greater than 65% when power is ramped down to 40 MW during non-peak hours. At a lower setpoint of 30 MW, the system could not maintain power setpoints and the HVS override was initiated. The setpoint of 30 MW is about as low as possible for modern-day power systems, which have been developed to operate at as low as 15% of capacity [53]. The results of that simulation still show a solar fraction approaching 70%, but at the expense of power control.

Further tests are carried out by adjusting the size of the storage and the field size to improve solar fraction while maintaining a power schedule with 30 MW as the off-peak setpoint. The tests follow a trial and error method where the TES and field are altered in size and the plant is tested in its ability to maintain power while achieving high solar fraction. This procedure is carried out until a solar fraction of 70% or more was realized with robust power control. First, the storage is increased in size to 8 h. This first test results in an operation that could maintain the drastic peaking schedule with the 30 MW setpoint and exhibits a solar fraction of 68%, slightly below the designated target. The field size is then increased to 740,000 m². This configuration results in a solar fraction of approximately 71%, while maintaining power control. The power production of June 24th can be seen in Figure 9 for a plant with a field size of 740,000 m² and 8-h storage. As can be seen, the marginal solar production of the plant is high when a drastic peaking schedule is implemented with TES large enough to handle excess solar energy from the larger collection field. Additionally, by increasing storage size, issues with stored energy dispatch appear to be alleviated somewhat, as can be seen by the less drastic dip between solar power production phases, which is likely due to longer periods of recycling at larger field sizes. At larger field sizes, the increased time of high incidence on the receiver results in recycling being active for a longer duration in the day. Thus, more time is used to charge the TES, which results in an increased exit temperature of the storage but one that remains under the maximum allowable recycling temperature. At smaller field sizes, the recycling does not occur as long, and that shorter time of charging cannot elevate the storage exit temperature.

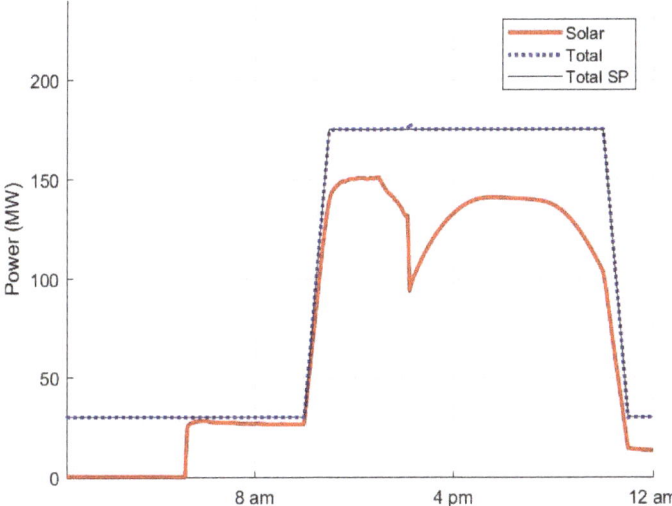

Figure 9. Power production of plant operating with storage bypass, extreme power peaking, and increased storage size to enhance solar fraction. The load is shown for June 24th.

4.4. Economic Evaluation of Recycling + Bypass Configuration

For the final plant configuration (recycle + bypass), an economic analysis gives insight to its viability compared to the base-case plant. To compare the two plants, the levelized cost of electricity (LCOE) is found following the peaking power schedule between 100 MW and 175 MW. The levelized cost of a power plant is the total cost over the plant's lifetime relative to the total amount of energy produced over that lifetime. When considering the yearly capital cost (C_t) and yearly operating cost (O_t), the LCOE of a plant is:

$$LCOE = \frac{\sum_{t=1}^{n} \left\{ \frac{C_t + O_t}{(1+r)^t} \right\}}{\sum_{t=1}^{n} E_{total,t}} \quad (23)$$

where t is the year, n is the lifetime for the plant, r is the discount rate for the plant over its lifetime and $E_{total,t}$ is the yearly energy production for year t. The yearly capital cost is found by converting the net present value of the total capital cost to annuity. Cost inflation is considered to estimate equipment costs for the year 2018 and for future yearly operation costs. The assumed parameters of the economic analysis are summarized in Table 3.

Table 3. Assumptions of the economic analysis.

Parameter	Value
Plant lifetime	25 years [54]
Inflation rate	4.5% [55]
Discount rate	5.5% [20]
Natural gas price	$6/MMBTU [54]

NREL's (National Renewable Emery Laboratory) cost model is used to estimate the cost of the plants' solar equipment, land, and general operation [54]. For any additional information needed, costs for the packed-bed TES and combined cycle equipment are approximated from literature [56,57]. A summary of capital and operating costs can be seen in Table 4 with the calculated LCOE for the base-case and recycle + bypass power plants.

Table 4. Economic analysis summary.

Description	Base-Case	Recycle + Bypass	Basis/Comments
Direct Capital Costs			
Heliostat field	$206,931,600	$255,824,700	$180/m^2 [54]
Tower + receiver	$71,890,300	$71,890,300	$105/kWt [54]
Packed bed TES	$14,310,200	$19,080,300	$10/kWht [56]
Combined cycle	$220,800,000	$220,800,000	$1104/kW [57]
Other Costs	$142,417,200	$146,174,900	Site improvements, plant balance, contingency [54]
Indirect Capital Costs			
EPC and Owner Cost	$74,350,700	$80,665,000	11% of direct capital costs [54]
Land Cost	$19,530,000	$19,530,000	$1953 acres at $10,000/acre [54]
Yearly Operating Costs			
Natural Gas	$34,441,900	$28,384,500	$6/MMBTU
Variable Operation	$3,526,700	$3,526,700	$4 MWh [54]
Fixed Operation	$11,122,200	$11,122,200	$51/kW-yr [54]
Total Lifetime Cost	$5,206,671,000	$4,899,579,700	
Yearly Energy Production	1,204,500 MWh	1,187,700 MWh	
LCOE	$172.9/MWh	$165.0/MWh	

The base-case and recycle + bypass plants exhibit a LCOE of $172.9/MWh and $165.0/MWh respectively, or roughly 4.6% reduction from the base-case to the recycle + bypass configuration. Despite having a larger field in the final configuration which leads to increased capital costs, that configuration is able to generate a larger portion of energy from solar activity. This higher solar fraction results in a reduction in natural gas usage and by extension reduced operation cost relative to the base-case plant. As a reference, the United States' Energy Information Administration (EIA) reports that as of 2018 the estimated LCOE for solar thermal plants is $165.1/MWh prior to tax credit [58]. This reported LCOE should be noted for solar thermal plants exhibiting an average capacity factor around 25%. For the plant presented, the capacity of the power plant is approximately 68.8% based upon the peaking schedule and 200 MW overall capacity. The capacity of reported solar-only CSP plants is restricted due to the limited amount of available solar energy and storage size. The hybrid plant proposed herein is able to leverage the hybrid operation to enhance the capacity and result in an LCOE similar to current CSP systems. Further LCOE reduction could be achieved by operating the plant at baseload throughout its lifetime, but high solar fractions would be sacrificed. Reduction of capital cost represents the major challenge in reducing the LCOE of CSP systems, with solar components representing 42% of overall capital cost for the final plant configuration. Cost reduction of these components may result in hybrid solar-combined cycle power plants being competitive with other power generation systems. However, the study presented herein shows that, through holistic design and operation, the LCOE of a hybrid CSP plant can be reduced for plants exhibiting very large solar fractions.

5. Conclusions

Investigation of a CSP-hybrid plant design and operation is undertaken to achieve very high solar fractions. A first-principles model of a tower-driven CSP hybridized with a combined cycle power plant is developed in Matlab/Simulink. The hybrid plant contains packed-bed thermal energy storage. The study presents a systematic approach where modifications are made to the configuration of the plant and the performance of each configuration is discussed in depth. Such an approach to achieve a high solar fraction has not been applied in CSP literature. Proposed plant configurations operate using a novel recycling configuration to control receiver exit temperature. Receiver exit temperature and power control are system performance metrics of particular interest. A base-case plant is shown to maintain power control and mitigation of temperatures at smaller field sizes. In cases with excessive solar energy available, represented here by increasing the field size in lieu of increased solar irradiance, the base-case cannot maintain a power setpoint, as the HVS-override must increase flow to maintain receiver temperature. The recycling operation is implemented to control temperature independently, resulting in greater reliability in power control. Additional flexibility observed in the proposed recycle scheme results in a yearly solar fraction over 30% when employing a peaking power load. This results in a 4.9% improvement in solar fraction relative to the base-case plant. At subsequently larger field sizes, the recycle control faces issues with inability to simultaneously control collector exit temperature and net power. Therefore, a bypass of storage is implemented to reduce storage charging rate and allow for longer recycle times. Introduction of a bypass allows for flexibility to install larger heliostat fields, which results in a higher solar fraction, tight control, and improved solar-to-electric efficiency. The system utilizing a bypass exhibits a 6% improvement in solar fraction when compared to the plant lacking a TES bypass. By implementing a drastic peaking power load schedule, solar fractions as high as 70% are realized for the final plant configuration. Lastly, an economic analysis shows that by implementing a recycling and TES bypass operation, the LCOE of such a power plant can be reduced by over 4% despite an increase in overall capital cost of the plant. To improve LCOE, further reduction of capital cost is necessary, specifically the capital cost of solar collection equipment. Operation of the plant at baseload can also lead to further reduced LCOE for the hybrid plant.

Author Contributions: Conceptualization, K.E. and K.P.; Methodology, K.E., K.R., and K.P.; Software, K.E. and K.P.; Validation, K.E. and K.P.; Formal Analysis, K.E., S.M.S., and K.R.; Investigation, K.E.; Resources, K.P.; Data Curation, K.E. and K.R.; Writing-Original Draft Preparation, K.E.; Writing-Review & Editing, K.E., S.M.S., and K.P.; Visualization, K.E., S.M.S., and K.P.; Supervision, K.P.; Project Administration, K.P.; Funding Acquisition, K.P.

Funding: This work is funded by United States' Department of Energy (DOE) under the DE-EE0007712 grant, which is affiliated with the DOE Industrial Assessment Center as well as further funding from the University of Utah Chemical Engineering Department.

Conflicts of Interest: The authors declare no conflict of interest.

Nomenclature

Acronym	Description
CSP	Concentrated solar power
CTR	Central tower receiver
DNI	Direct normal irradiance
STE	Solar-to-electric efficiency
SF	Solar fraction
ISCC	Integrate solar combined cycle
IGV	Inlet guide vane
NTU	Number of Heat Transfer Units
NREL	National Renewable Energy Laboratory
PTC	Parabolic trough collector
APC	Advanced process control
TES	Thermal energy storage
HVS	High-value selector
HRSG	Heat recovery steam generator
LCOE	Levelized cost of electricity

Symbol	Description	Value	Units
$\dot{m}_{a,n}$	Nominal baseload air flow rate	537	kg/s
$\dot{m}_a$	Air-flow rate	-	kg/s
$\dot{m}_{f,n}$	Nominal baseload fuel flow rate	10.2	kg/s
$\dot{m}_f$	Fuel flow rate	-	kg/s
h_i	Internal air flow heat transfer coefficient for CTR	-	kW/m²·K
h_o	External air flow heat transfer coefficient for CTR	-	kW/m²·K
h_s	Internal air flow heat transfer coefficient for TES	-	kW/m²·K
A_c	Heat transfer area glass	300	m²
A_p	Heat transfer area receiver pipe (per pipe)	3.73	m²
A_s	Heat transfer area of the stone medium in TES	-	m²
$P_{amb,o}$	Ambient atmospheric reference pressure	1	atm
P_{amb}	Ambient atmospheric pressure	-	atm
T_a	Temperature of air	-	K
$T_{amb,o}$	Ambient atmospheric reference temperature	288.15	K
T_{amb}	Ambient atmospheric temperature	-	K
T_c	Temperature of receiver glass	-	K
T_d	Compressor outlet temperature	-	K
T_e	Turbine exhaust temperature	-	K
T_f	Turbine firing temperature setpoint	1396	K
ΔT_g	Temperature change of flue gas in HRSG	-	K
T_p	Temperature of receiver pipe	-	K
V_a	Shell volume of flowing air	-	m³
V_c	Shell volume of receiver glass	-	m³
V_p	Shell volume of receiver pipe	-	m³
$c_{p,h}$	Specific heat of combustion exhaust gas	1.157	kJ/kg·K
$c_{p,a}$	Heat capacity of air	-	kJ/kg·K
$c_{p,c}$	Heat capacity of receiver glass	840	kJ/kg·K
$c_{p,p}$	Heat capacity of receiver pipe	0.574	kJ/kg·K
$c_{p,s}$	Heat capacity of TES	-	kJ/kg·K
q_{inc}	Incident concentrated solar irradiance	-	kW/m²
γ_h	Hot end ratio of specific heats	1.33	-
γ_c	Cold end ratio of specific heats	1.4	-
ε_c	Emissivity of receiver glass	0.9	-
ε_p	Emissivity of receiver pipe	0.25	-
η_c	Compressor efficiency	86	%
η_{comb}	Combustion efficiency	99	%
η_f	Nominal fuel to electric efficiency	-	%
η	Overall plant efficiency	-	%
θ_{IGV}	IGV angle	-	°
θ_{max}	Maximum IGV angle	85.0	°
θ_{min}	Minimum IGV angle	11.6	°
ν_p	Absorptivity of receiver pipe	0.97	-
ρ_a	Density of air	-	kg/m³
ρ_c	Density of receiver glass	2400	kg/m³
ρ_p	Density of receiver pipe	7850	kg/m³
ρ_s	Density of TES medium	1933	kg/m³
τ_c	Transmissivity of receiver glass	0.96	-
LHV	Lower heating value of fuel	46,000	kJ/kg
PR_C	Compression ratio of compressor	15.4	-
PR_T	Compression ratio of turbine	15.4	-
SF	Solar fraction	-	%
STE	Solar-to-electric efficiency	-	%

Symbol	Description	Value	Units
σ	Stefan-Boltzmann constant	5.67×10^{-8}	W/m$^2 \cdot$K^4
ϵ	Effectiveness of HRSG	0.80	-
C_{min}	Minimum heat capacity rate	-	kW/K
ΔT_g	Temperature drop of flue gas of HRSG	-	K
Δh_j	Enthalpy change of steam/water across unit j	-	kJ/kg
$\dot{m}_w$	Flow rate of steam/water	-	kg/s
Q_i	Heat rate of component i (HRSG or condenser)	-	kW
W_i	Work of component i (steam turbine or pump)	-	kW
$Power_{net}$	Net plant power production	-	MW
A_{field}	Heliostat field total incident area	-	m^2

References

1. Zhang, H.L.; Baeyens, J.; Eve, J.D.; Eres, G.C. Concentrated solar power plants: Review and design methodology. *Renew. Sustain. Energy Rev.* **2013**, *22*, 466–481. [CrossRef]
2. Kuravi, S.; Trahan, J.; Goswami, D.Y.; Rahman, M.M.; Stefanakos, E.K. Thermal energy storage technologies and systems for concentrating solar power plants. *Prog. Energy Combust. Sci.* **2013**, *39*, 285–319. [CrossRef]
3. Madaeni, S.H.; Member, S.; Sioshansi, R.; Denholm, P. How Thermal Energy Storage Enhances the Economic Viability of Concentrating Solar Power. *Proc. IEEE* **2012**, *10*, 335–347. [CrossRef]
4. Singh, H.; Saini, R.P.; Saini, J.S. A review on packed bed solar energy storage systems. *Renew. Sustain. Energy Rev.* **2010**, *14*, 1059–1069. [CrossRef]
5. Grange, B.; Dalet, C.; Falcoz, Q.; Siros, F.; Ferrière, A. Simulation of a Hybrid Solar Gas-turbine Cycle with Storage Integration. *Energy Procedia* **2014**, *49*, 1147–1156. [CrossRef]
6. Grange, B.; Dalet, C.; Falcoz, Q.; Ferriere, A.; Flamant, G. Impact of thermal energy storage integration on the performance of a hybrid solar gas-turbine power plant. *Appl. Therm. Eng.* **2016**, *105*, 266–275. [CrossRef]
7. Kalogirou, S. *Solar Energy Engineering Processes and Systems*; Academic Press: Cambridge, MA, USA, 2009.
8. Powell, K.M.; Edgar, T.F. Modeling and control of a solar thermal power plant with thermal energy storage. *Chem. Eng. Sci.* **2012**, *71*, 138–145. [CrossRef]
9. Alva, G.; Lin, Y.; Fang, G. An overview of thermal energy storage systems. *Energy* **2018**, *144*, 341–378. [CrossRef]
10. Barlev, D.; Vidu, R.; Stroeve, P. Innovation in concentrated solar power. *Sol. Energy Mater. Sol. Cells* **2011**, *95*, 2703–2725. [CrossRef]
11. Pelay, U.; Luo, L.; Fan, Y.; Stitou, D.; Rood, M. Thermal energy storage systems for concentrated solar power plants. *Renew. Sustain. Energy Rev.* **2017**, *79*, 82–100. [CrossRef]
12. Johnson, E.; Bates, L.; Dower, A.; Bueno, P.C.; Anderson, R. Thermal energy storage with supercritical carbon dioxide in a packed bed: Modeling charge-discharge cycles. *J. Supercrit. Fluids* **2018**, *137*, 57–65. [CrossRef]
13. Klein, P.; Roos, T.; Sheer, T. Parametric analysis of a high temperature packed bed thermal storage design for a solar gas turbine. *Sol. Energy* **2015**, *118*, 59–73. [CrossRef]
14. Tian, Y.; Zhao, C.Y. A review of solar collectors and thermal energy storage in solar thermal applications. *Appl. Energy* **2013**, *104*, 538–553. [CrossRef]
15. Ju, X.; Wei, G.; Du, X.; Yang, Y. A novel hybrid storage system integrating a packed-bed thermocline tank and a two-tank storage system for concentrating solar power (CSP) plants. *Appl. Therm. Eng.* **2016**, *92*, 24–31. [CrossRef]
16. Barigozzi, G.; Bonetti, G.; Franchini, G.; Perdichizzi, A.; Ravelli, S. Thermal performance prediction of a solar hybrid gas turbine. *Sol. Energy* **2012**, *86*, 2116–2127. [CrossRef]
17. Peterseim, J.H.; Tadros, A.; White, S.; Hellwig, U.; Landler, J.; Galang, K. Solar Tower-biomass Hybrid Plants—Maximizing Plant Performance. *Energy Procedia* **2014**, *49*, 1197–1206. [CrossRef]
18. Santos, M.J.; Merchán, R.P.; Medina, A.; Hernández, A.C. Seasonal thermodynamic prediction of the performance of a hybrid solar gas-turbine power plant. *Energy Convers. Manag.* **2016**, *115*, 89–102. [CrossRef]
19. Powell, K.M.; Rashid, K.; Ellingwood, K.; Tuttle, J.; Iverson, B.D. Hybrid concentrated solar thermal power systems: A review. *Renew. Sustain. Energy Rev.* **2017**, *80*, 215–237. [CrossRef]

20. Rashid, K.; Safdarnejad, S.M.; Powell, K.M. Dynamic simulation, control, and performance evaluation of a synergistic solar and natural gas hybrid power plant. *Energy Convers. Manag.* **2019**, *179*, 270–285. [CrossRef]
21. Powell, K.; Hedengren, J.; Hedengren, J.D.; Edgar, T.F. Dynamic Optimization of a Solar Thermal Energy Storage System over a 24 Hour Period using Weather Forecasts. In Proceedings of the 2013 American Control Conference, Washington, DC, USA, 17–19 June 2013.
22. Çengel, Y.A.; Boles, M.A. *Thermodynamics: An Engineering Approach*; McGraw-Hill Education: New York, NY, USA, 2014.
23. Bonadies, M.F.; Mohagheghi, M.; Ricklick, M.; Kapat, J.S. Solar retrofit to combined cycle power plant with thermal energy storage. In Proceedings of the ASME Turbo Expo, Glasgow, UK, 14–18 June 2010; pp. 921–931.
24. Kribus, A.; Zaibel, R.; Carey, D.; Segal, A.; Karni, J. A solar-driven combined cycle power plant. *Sol. Energy* **1998**, *62*, 121–129. [CrossRef]
25. Spelling, J.; Favrat, D.; Martin, A.; Augsburger, G. Thermoeconomic optimization of a combined-cycle solar tower power plant. *Energy* **2012**, *41*, 113–120. [CrossRef]
26. Heller, P.; Pfaender, M.; Denk, T. Test and evaluation of a solar gas turbine system. *Sol. Energy* **2006**, *80*, 1225–1230. [CrossRef]
27. Korzynietz, R.; Brioso, J.A.; Del Río, A.; Quero, M.; Gallas, M.; Uhlig, R.; Ebert, M.; Buck, R.; Teraji, D. Solugas—Comprehensive analysis of the solar hybrid Brayton plant. *Sol. Energy* **2016**, *135*, 578–589. [CrossRef]
28. Quero, M.; Korzynietz, R.; Ebert, M.; Jiménez, A.A.; del Río, A.; Brioso, J.A. Solugas—Operation experience of the first solar hybrid gas turbine system at MW scale. *Energy Procedia* **2013**, *49*, 1820–1830. [CrossRef]
29. Olivenza-León, D.; Medina, A.; Hernández, A.C. Thermodynamic modeling of a hybrid solar gas-turbine power plant. *Energy Convers. Manag.* **2015**, *93*, 435–447. [CrossRef]
30. Behar, O.; Khellaf, A.; Mohammedi, K. A review of studies on central receiver solar thermal power plants. *Renew. Sustain. Energy Rev.* **2013**, *23*, 12–39. [CrossRef]
31. Okoroigwe, E.; Madhlopa, A. An integrated combined cycle system driven by a solar tower: A review. *Renew. Sustain. Energy Rev.* **2016**, *57*, 337–350. [CrossRef]
32. Camacho, E.F.; Gallego, A.J. Optimal operation in solar trough plants: A case study. *Sol. Energy* **2013**, *95*, 106–117. [CrossRef]
33. Pasamontes, M.; Álvarez, J.D.; Guzmán, J.L.; Lemos, J.M.; Berenguel, M. A switching control strategy applied to a solar collector field. *Control Eng. Pract.* **2011**, *19*, 135–145. [CrossRef]
34. Camacho, E.F.; Berenguel, M.; Gallego, A.J. Control of thermal solar energy plants. *J. Process Control* **2014**, *24*, 332–340. [CrossRef]
35. Camacho, E.F.; Gallego, A.J. Model Predictive Control In Solar Trough Plants: A Review. *IFAC-PapersOnLine* **2015**, *48*, 278–285. [CrossRef]
36. Silva, R.N.; Rato, L.M.; Lemos, J.M.; Coito, F. Cascade control of a distributed collector solar field. *J. Process Control* **1997**, *7*, 111–117. [CrossRef]
37. Camacho, E.F.; Berenguel, M.; Alvarado, I.; Limon, D. Control of Solar Power Systems: A survey. *IFAC Proc. Vol.* **2010**, *43*, 817–822. [CrossRef]
38. Camacho, E.F.; Berenguel, M. Control of Solar Energy Systems. *IFAC Proc. Vol.* **2012**, *45*, 848–855. [CrossRef]
39. Powell, K.M.; Edgar, T.F. Control of a large scale solar thermal energy storage system. In Proceedings of the 2011 American Control Conference, San Francisco, CA, USA, 29 June–1 July 2011; pp. 1530–1535.
40. Juuso, E.K.; Yebra, L.J. Smart adaptive control of a solar collector field. *IFAC Proc. Vol.* **2014**, *47*, 2564–2569. [CrossRef]
41. Valdés, M.; Rapún, J.L. Optimization of heat recovery steam generators for combined cycle gas turbine power plants. *Appl. Therm. Eng.* **2001**, *21*, 1149–1159. [CrossRef]
42. Beasley, D.E.; Clark, J.A. Transient response of a packed bed for thermal energy storage. *Int. J. Heat Mass Transf.* **1984**, *21*, 1659–1669. [CrossRef]
43. Ellingwood, K.; Tuttle, J.; Powell, K. Leveraging storage and hybridization to maximize renewable utilization. In Proceedings of the AIChE Annual Meeting, San Francisco, CA, USA, 13–18 November 2016.
44. Kim, J.S.; Powell, K.M.; Edgar, T.F. Nonlinear model predictive control for a heavy-duty gas turbine power plant. In Proceedings of the 2013 American Control Conference, Washington, DC, USA, 17–19 June 2013; pp. 2952–2957.
45. MathWorks Inc. *MATLAB and Statistics Toolbox Release 2017a*; MathWorks Inc.: Natick, MA, USA, 2017.

46. Cengel, Y.; Cimbala, J.; Turner, R. *Fundementals of Thermal-Fluid Sciences*, 4th ed.; McGraw-Hill: New York, NY, USA, 2012.
47. Ramsey, J.; Kuehn, T. Appendix D: Solar Radiation. 2018. Available online: http://www.me.umn.edu/courses/me4131/LabManual/AppDSolarRadiation.pdf (accessed on 1 May 2017).
48. Wilkes, J.O. *Fluid Mechanics for Chemical Engineers*, 2nd ed.; Pearson Education, Inc.: London, UK, 2006.
49. Incropera, F.P.; Dewitt, D.P.; Bergman, T.L.; Lavine, A.S. *Fundamentals of Heat and Mass Transfer*; John Wiley: Hoboken, NJ, USA, 2007.
50. Spirax Sarco. Pages—Steam Tables. 2018. Available online: http://www.spiraxsarco.com/Resources/Pages/steam-tables.aspx (accessed on 9 February 2018).
51. NREL. NSRDB Update—TMY3: Alphabetical List by State and City. Available online: http://rredc.nrel.gov/solar/old_data/nsrdb/1991-2005/tmy3/by_state_and_city.html (accessed on 10 May 2017).
52. Peng, S.; Wang, Z.; Hong, H.; Xu, D.; Jin, H. Exergy evaluation of a typical 330 MW solar-hybrid coal-fired power plant in China. *Energy Convers. Manag.* **2014**, *85*, 848–855. [CrossRef]
53. Ansolda Energia. *Innovation Based on Proven Technology*; Ansolda Energia: Genova, Italy, 2018.
54. Turchi, C.S.; Heath, G.A. *Molten Salt Power Tower Cost Model for the System Advisor Model (SAM)*; National Renewable Energy Lab: Golden, CO, USA, 2013.
55. Mansouri, M.T.; Ahmadi, P.; Kaviri, A.G.; Jaafar, M.N.M. Exergetic and economic evaluation of the effect of HRSG configurations on the performance of combined cycle power plants. *Energy Convers. Manag.* **2012**, *58*, 47–58. [CrossRef]
56. Allen, K.; von Backström, T.; Joubert, E.; Gauché, P. Rock bed thermal storage: Concepts and costs. *AIP Conf. Proc.* **2016**, *1734*, 50003.
57. EIA. *Capital Cost Estimates for Utility Scale Electricity Generating Plants*; EIA: Washington, DC, USA, 2016.
58. EIA. *Levelized Cost and Levelized Avoided Cost of New Generation Resources in the Annual Energy Outlook 2018*; EIA: Washington, DC, USA, 2018.

© 2018 by the authors. Licensee MDPI, Basel, Switzerland. This article is an open access article distributed under the terms and conditions of the Creative Commons Attribution (CC BY) license (http://creativecommons.org/licenses/by/4.0/).

Article

Optimized Dimensioning and Operation Automation for a Solar-Combi System for Indoor Space Heating. A Case Study for a School Building in Crete

Dimitris Al. Katsaprakakis * and Georgios Zidianakis

Wind Energy and Power Plants Synthesis Laboratory, Department of Mechanical Engineering, Technological Educational Institute of Crete, Estavromenos, 714 10 Heraklion Crete, Greece; g.zidianakis82@gmail.com
* Correspondence: dkatsap@staff.teicrete.gr; Tel.: +30-2810-379-220; Fax: +30-2810-319-478

Received: 30 November 2018; Accepted: 3 January 2019; Published: 7 January 2019

Abstract: This article investigates the introduction of hybrid power plants for thermal energy production for the indoor space heating loads coverage. The plant consists of flat plate solar collectors with selective coating, water tanks as thermal energy storage and a biomass heater. A new operation algorithm is applied, maximizing the exploitation of the available thermal energy storage capacity and, eventually, the thermal power production from the solar collectors. An automation system is also designed and proposed for the realization of the newly introduced algorithm. The solar-combi system is computationally simulated, using annual time series of average hourly steps. A dimensioning optimization process is proposed, using as criterion the minimization of the thermal energy production levelized cost. The overall approach is validated on a school building with 1000 m^2 of covered area, located in the hinterland of the island of Crete. It is seen that, given the high available solar radiation in the specific area, the proposed solar-combi system can guarantee the 100% annual heating load coverage of the examined building, with an annual contribution from the solar collectors higher than 45%. The annually average thermal power production levelized cost is calculated at 0.15 €/kWh$_{th}$.

Keywords: heating and cooling loads; biomass-solar combi systems; buildings energy performance upgrade; solar collectors' simulation; thermal energy storage

1. Introduction

1.1. Solar Radiation for Thermal Energy Production

Thermal energy constitutes one of the major consumed final forms of energy, accounting for a high percentage of the annual energy consumption balance of a specific consumer. For instance, the thermal energy annual consumption for indoor space conditioning and hot water production in a typical commercial building in United States accounts for more than 40% of the building's total energy consumption [1].

Thermal energy so far is traditionally produced with electricity or non-renewable fossil fuels (oil, natural gas, biomass etc.). Thermal energy production with electrical devices-maybe apart from heat pumps with Coefficient of Performance higher than 5 constitutes a considerably ineffective process, especially when electricity is produced from thermal power plants. In such cases, the overall efficiency of the whole energy transformation process, starting from the initial chemical primary energy source of the available fossil fuel and ending in the finally consumed thermal energy, can be close to 35%, depending on the thermal generators types involved. This is a typical average efficiency of electrical systems based on thermal power plants, like the one in Crete, although new advances on the thermal generators technologies promise higher efficiencies in the approximate future. On the other hand, thermal energy produced locally with oil, gas or solid fuels burners (e.g., coal or biomass), imposes significant gas emissions and atmospheric pollution, which can raise a crucial issue in cases of already burdened urban environments.

Thermal energy production technologies based on Renewable Energy Sources (RES) features as a quite attractive alternative, especially in southern climates (e.g., the Mediterranean basin), with high incident solar radiation density, even during winter. The available thermal energy production technologies from RES include the various types of solar collectors: uncovered, flat plate with/without selective coating, vacuum tubes and concentrating solar collectors.

Concentrating solar collectors, due to the high achieved concentration ratio of the incident solar radiation on the focal line or point, and the subsequent high concentrated thermal power, are used in the so-called solar thermal power plants, aiming at guaranteed electricity production from solar radiation [2,3].

The uncovered solar collectors consist of a set of plastic, dark colored pipelines (usually black). The most usual application of uncovered solar collectors is the swimming pools heating. With uncovered solar collectors and required water temperature in the swimming pool at the range of 25 °C, the use of a swimming pool in warm climates (annual global irradiation on horizontal level higher than 1600 kWh/m^2) can be extended from April to October. Empirically, the total required solar collectors' area is equal to 80% of the heated swimming pool's free surface [4–6].

Flat-plate solar collectors are the most widely used solar radiation exploitation technology for the production of thermal energy. The absorptance coefficient of the absorber plate determines the type of the flat-plate solar collectors: black painted or semi-selective coating collectors, with absorptance coefficients at the range of 80%, and selective coating collectors, with absorptance coefficient at 90–95%, while keeping the emissivity coefficient at 5–15%. The flat-plate solar collectors are used for indoor space and water heating [7,8]. Selective coating collectors may be also used when higher temperatures are required [9,10].

Vacuum tube solar collectors consist of an absorber surface, introduced inside a vacuum space, aiming at the minimization of the thermal losses with convection from the collector to the ambient. Vacuum tube solar collectors can maintain high efficiency even during adverse weather conditions. They seem ideal for indoor space and water heating in cold climates [11–15]. Additionally, they exhibit the ability to increase significantly the temperature of the working medium (up to 300 °C). Consequently, they are used in special industrial applications and generally, in every application that high temperatures are required [16,17].

Finally, a very recent innovation concerning the exploitation of solar radiation are the photovoltaic thermal hybrid solar collectors. These devices exploit solar radiation for both electricity production, acting as a common photovoltaic panel, and for thermal energy production, acting as a flat plate solar collector, cooling, concurrently, through the flow of the working medium in the pipelines, the photovoltaic panel and improving, in this way, its efficiency [18–20]. The total efficiency for both thermal energy and electricity production is increased above 85%, calculated as the ratio of the total electrical and thermal power output versus the incident solar radiation on the total photovoltaic hybrid collector's effective surface.

1.2. Solar-Combi Systems

The combination of solar collectors with thermal energy storage tanks and a conventional back-up thermal power production unit (e.g., an oil burner) is known as solar-combi system. In other words, this implementation could be defined as a hybrid power plant for thermal power production. The essential layout of a hybrid thermal power plant is presented in Figure 1. It consists of the following discrete components:

- solar collectors, as base units
- water thermal tanks, as storage units
- a central heating burner, as the back-up unit for guaranteed thermal power production

- an electronic central control unit, for the supervision and the management of the system's operation
- the hydraulic network, consisting of pipelines, devices and equipment aiming to ensure the secure and effective circulation of the working fluid.

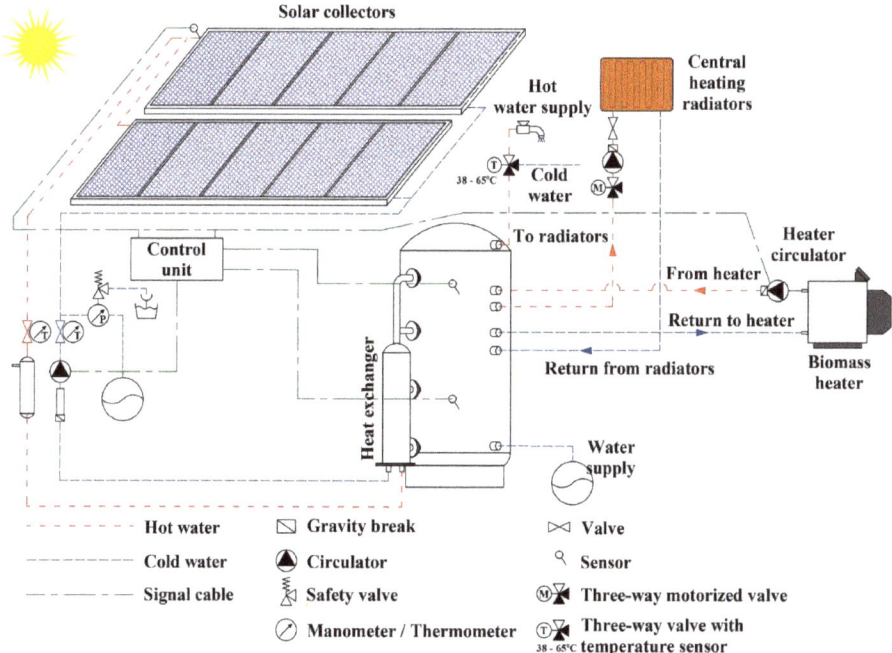

Figure 1. Typical layout of a thermal hybrid power plant.

As with hybrid power plants for electricity production, the ultimate scope of solar-combi systems is the guaranteed coverage of a specific thermal power demand, based on solar collectors, namely on the stochastic availability of solar radiation (primary energy source). This is achieved with the support of thermal storage tanks, which, in most conventional cases are insulated water tanks, equipped with multiple inlets for the concurrent thermal energy storage from different hydraulic networks, supplied by alternative thermal sources. The system is integrated with a conventional thermal power production unit (e.g., an oil burner), acting exclusively as a back-up unit, namely it is utilized only in cases of inadequate available solar radiation and low thermal energy storage level, with regard to the current thermal power demand.

Solar-combi systems constitute a popular subject in the relevant scientific literature. They are usually introduced and investigated for indoor space heating for residential buildings [21–25], while studies have been also executed for school buildings [26], hotels [27], swimming pools heating [28,29], even for central heating district systems [30] and advanced systems for both pure and hot water production [31]. The relevant research has been concentrated on:

- the simulation of their operation, based on specific operation algorithms [32,33]
- validation of their performance based on the comparison of both a simulation approach and experimental measurements of the system's operation [34]
- investigation of alternative thermal energy storage technologies, involving phase change materials [35,36]
- evaluation of solar-combi systems layouts involving seasonal thermal storage [23,37,38]
- evaluation of solar-combi systems performance under different operation conditions [39].

1.3. Content, Scope and Novelties of the Present Article

The scope of this article is the optimum dimensioning of a solar-combi system for indoor space heating and the introduction of a new operation algorithm, along with its realization system. The overall approach is validated on a high-school building, located in the small town of Arkalochori (5000 permanent inhabitants), in the hinterland of Crete, where remarkable solar radiation is available even during the winter period. Particularly, the study focuses on the following tasks:

- calculation of the building's heating loads
- dimensioning of the solar-combi system, consisting of flat plate solar collectors with selective coating, water thermal energy storage tanks and a biomass heater, aiming at 100% coverage of the building's heating loads
- annual calculation of the thermal energy production and storage from the involved components of the solar-combi system
- optimization of the system's dimensioning, using as criterion the minimisation of the average annual thermal energy production levelized cost (in €/kWh$_{th}$). This cost is calculated as the ratio of the total average annual production cost (including regular annual operation & maintenance cost and the set-up cost annual amortization) versus the final thermal energy production by the solar-combi system. A detailed definition is provided in relationship (1), Section 3.1.

The above tasks are executed with the support of the TRNSYS software application, regarding the heating loads calculation and the simulation of the solar collectors' performance. The combined operation of the solar collectors, the thermal storage tanks and the biomass heater is executed with simulation applications of the introduced operation algorithm, developed by the authors on the LabVIEW platform. The calculation is executed on hourly average calculation steps and for one whole annual period.

The main challenge with the proposed system was that, due to the size of the building under consideration, the required capacity of the thermal storage facility is considerably increased. The increased thermal energy storage capacity can be met only with the introduction of multiple water thermal storage tanks. Yet, this fact makes the control and the automatic operation of the solar-combi system more complicated, since there should be a flexible and effective automation introduced for the realization of an optimized storage algorithm in the available different storage tanks in order to:

(a) ensure that there will always be hot water available at the appropriate temperature inside a specific thermal tank, to undertake the current thermal power demand.
(b) maximize the thermal energy storage from the solar collectors, even in cases it is provided in relatively low temperatures (e.g., during the very first hours, right after the dawn).

The novelty of this article lays precisely on the development of the operation algorithm of the investigated solar-combi system, configured in order to fulfil the above objectives, and the control system introduced for its realization. The realization process and the corresponding results of the above tasks are presented in the following sections.

2. Heating Loads Calculation

2.1. Location, Background

The high-school building under consideration (Figure 2) is located in the small town of Arkalochori, in the hinterland of the island of Crete, 25 km to the south from the city of Heraklion, the capital of the island. The geographical coordinates of the building are 35°09′20″ N, 25°16′08″ E. The energy upgrade of the school building from energy performance rank D to B+ was funded by a national—European Union (E.U.) co-funding action, following an open tender posted by the responsible Ministry of Energy in 2011.

Figure 2. General view of the examined school building.

The annual energy sources recorded consumptions, based on the corresponding statements and invoices from the providers, before the building's energy upgrade were:

- electricity: 23,351 kWh for indoor and outdoor spaces lighting and for offices and laboratories devices
- diesel oil: 4500 lt, exclusively for the indoor space heating.

The building was adequately insulated, following the Hellenic Directive on Buildings' Energy Performance, valid in 2011, given also the relatively mild winter and cool summer, typical features in Mediterranean climates. No passive measures were applied. The energy upgrade of the building was mainly based on:

- the replacement of the old, energy consuming bulbs with new ones with LED technology
- the introduction of roof fans for physical cooling at the end and the beginning of the academic season
- the replacement of the existing diesel oil heater with a new solar-combi system.

This article will present in detail the introduced solar-combi system.

2.2. Climate Conditions—R.E.S. Potential

The climate at the location of the examined school building is typical Mediterranean. It is characterized by mild winters, with temperatures from 5 to 17 °C, and relatively cool summers with temperatures rarely higher than 30–32 °C and relative humidity below 75% during summer, due to the prevailing local north-west winds.

The incident solar radiation on the horizontal surface is presented in Figure 3, measured by a local meteorological station installed by the Hellenic National Meteorological Service (HNMS) [40], in a location roughly 15 km away from the building' location (coordinates 35°12′30″ N, 25°20′20″ E). In Figure 3 is shown that the solar radiation can exceed 900 W/m^2 during summer, with a yearly global irradiation at 1857 kWh/m^2. During winter it normally reaches values above 500 W/m^2.

In Figure 4 the annual time series of the wind velocity is presented, as measured by the same meteorological station. The annual average wind velocity is measured at 3.58 m/s, while the Weibull parameters are calculated at C = 3.99 m/s and k = 1.42. The above features indicate the existence of relatively mild wind conditions in the examined location. This annual wind velocity time series will be employed in this work for the calculation of:

- the thermal losses rate from the solar collectors
- the heating and cooling loads of the indoor space.

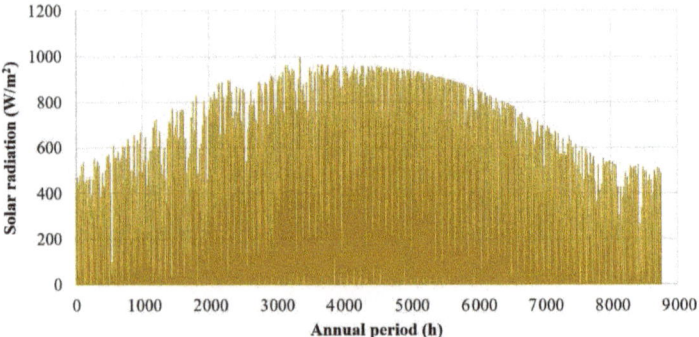

Figure 3. The annual time series of the available incident solar radiation on horizontal plane at the town of Arkalochori.

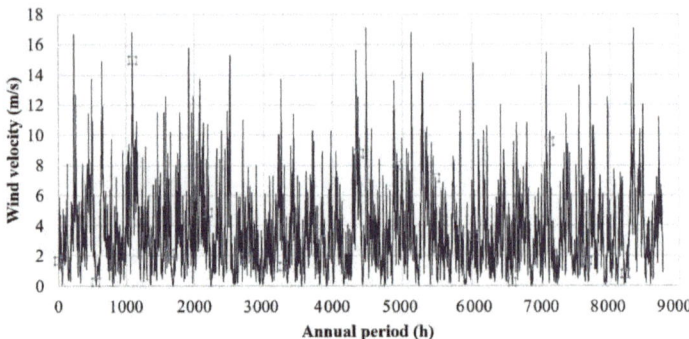

Figure 4. The annual wind velocity time series as measured by the meteorological station of the HNMS.

In Figure 5 the annual time series of the ambient temperature is presented, as measured by the above mentioned meteorological mast. The annual temperature and solar radiation time series will be employed in this work for the calculation of:

- the thermal power production from the introduced solar collectors
- the heating and cooling loads of the indoor space.

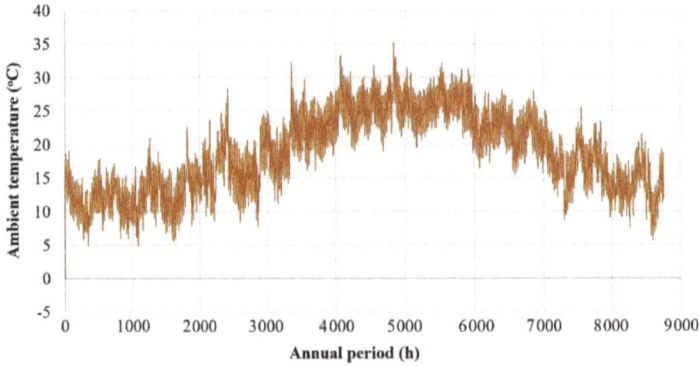

Figure 5. The annual ambient temperature time series at the town of Arkalochori.

2.3. Heating and Cooling Loads Calculation

The building is extended on three floors (basement, ground and first floor). The covered area per floor and the determination of the thermal zones are presented in Table 1. Thermal zone 1 includes all the conditioned indoor space: classrooms, teachers' offices, laboratories, the events' hall and the toilets. Thermal zone 2 includes all the non-conditioned spaces in the building: escalators, repositories and boiler—machinery room. The constructive features of the building's envelope with their U-factors are presented in Table 2. The solar gain factor of the transparent surfaces is set at 0.77.

Table 1. Covered areas and volumes of the conditioned and non-conditioned indoor space in the examined school building.

Floor	Thermal Zone 1 (Conditioned Spaces)		Thermal Zone 2 (Non-Conditioned Spaces)	
	Area (m^2)	Volume (m^3)	Area (m^2)	Volume (m^3)
Basement	31	108	368	1282
Ground floor	633	2216	347	1215
First floor	382	1337	133	471
Total	1046	3661	848	2968

Table 2. Constructive features and U-factors of the building's envelope.

Constructive Element	Description	U-Factor (W/m^{2}K)
Basement ground	marble-lime plaster-insulation layer-water sealing sheet-reinforced concrete plate	0.985
Floors' ground	marble-lime plaster-reinforced concrete plate-plaster-paint coating	2.985
Roof	concrete plates-lime plaster-elastic asphalted cardboard-insulation layer-reinforced concrete plate-plaster-paint coating	1.050
Vertical external walls	paint coating-plaster-bricks-expanded polystyrene-bricks-plaster-paint coating	1.055
Internal vertical walls	paint coating-plaster-bricks-plaster-paint coating	2.125
Windows—Doors	aluminum frame, with no thermal brake, double 4 mm glazing with 6 mm gap, no reflective coating	2.70

The calculation of the heating and cooling loads is based on the essential relevant methodology [41] and was executed with TRNSYS. The building's simulation model and the logical diagram introduced in the software application are presented in Figure 6.

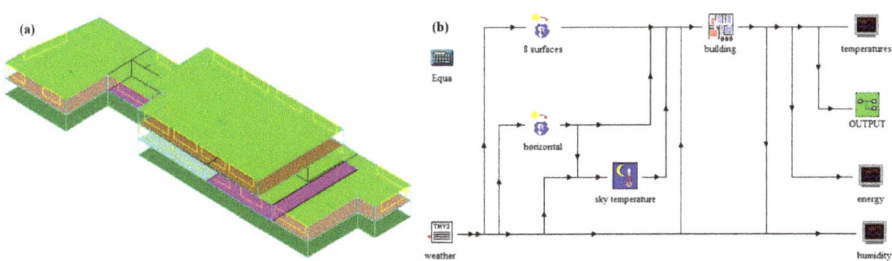

Figure 6. Building's computational simulation model (**a**) and calculation logical diagram (**b**) introduced in TRNSYS for the heating and cooling loads calculation.

The involved thermophysical parameters are introduced either from the Hellenic Directive on Buildings' Energy Performance [42] or the ASHRAE Fundamentals [43]. Specifically:

- the thermal transition coefficients h_i and h_o for the indoor or the ambient space respectively to the envelope were set:

- $h_i = 10$ W/m²K and $h_o = 25$ W/m²K for air flow over horizontal surfaces and for an average wind speed of 5 m/s
- $h_i = 7.7$ W/m²K and $h_o = 25$ W/m²K for air flow next to vertical surfaces and for an average wind speed of 5 m/s

- the natural ventilation coefficient from the openings' frames is set equal to 1
- thermal comfort conditions: temperature 22 °C in winter and 26 °C in summer, relative humidity 50%
- ventilation requirements: 26 m³/h & user [42]
- internal heat gains from humans, devices etc., as defined in the relevant Hellenic Directive on Buildings' Energy Performance [42]
- the daily operation schedule and the average users number were provided by the building's Management: Monday–Friday: 8:00–17:00, from the 10th of September to the 15th of June; from the 16th of June to the 10th of September school remains closed (only the offices operate for the remaining days of September and June).

Table 3. Monthly analysis of the heating and cooling loads of the examined building.

Months	Monthly Total Thermal Loads (kWh)		Monthly Peak Loads (kW)	
	Heating	Cooling	Heating	Cooling
January	8243	23	240.94	1.66
February	7026	60	219.70	2.65
March	5946	160	198.86	3.40
April	841	409	128.36	60.14
May	0	1546	80.08	80.98
June	0	1820	5.93	76.64
July	0	0	0.00	0.00
August	0	0	0.00	0.00
September	0	1656	3.51	80.48
October	422	622	81.06	61.71
November	3316	6	158.11	0.95
December	6561	11	214.88	1.29
Totals/max	32,356	6314	240.94	80.98
Total heating and cooling	35,670		-	-
Totals specific	30.93	6.04	-	-

In Table 3 the monthly summarized results from the heating and cooling loads calculation are presented. The results presented in this table can be summarized as follows:

- The annual final thermal energy consumed for heating and cooling is calculated at 32.4 MWh and 6.3 MWh, respectively.
- The total final specific thermal energy consumption per unit of conditioned space covered area (1046 m²) is calculated equal to 30.93 kWh/m² for heating and 6.04 kWh/m² for cooling.
- The total final thermal energy specific consumption for the conditioning of the indoor space is calculated at 36.97 kWh/m².

By assuming the following efficiencies of the central heating system:

- diesel oil heater: 0.80
- heating distribution hydraulic network: 0.88
- heating radiators: 0.92

and the diesel oil lowest calorific value equal to 10.25 kWh/lt [42], it is calculated that the annual oil consumption for the 100% heating loads coverage of the school building should be:

$$32{,}356 \text{ kWh}/(0.80 \times 0.88 \times 0.92 \times 10.25 \text{ kWh/lt}) = 4874 \text{ lt}$$

By comparing the above result with the recorded annual diesel oil consumption (4500 lt), the accuracy and the adequacy of the computational simulation process applied for the calculation of the building's heating loads is verified.

By introducing the diesel oil procurement price in Crete at 0.95 €/lt, the annual diesel oil procurement cost for the total annual coverage of the building's heating loads is calculated at 4630 €. The thermal power production annual cost, accounting:

- the above calculated diesel oil procurement cost
- an annual maintenance cost of 200 €
- the annual amortization of the invested capital assumed at 5000 € over a period of 20 years

is calculated at 5080 €. The corresponding levelized production cost is calculated at 0.1570 €/kWh$_{th}$.

In Figure 7 the annual variation of the school building's heating and cooling loads is depicted.

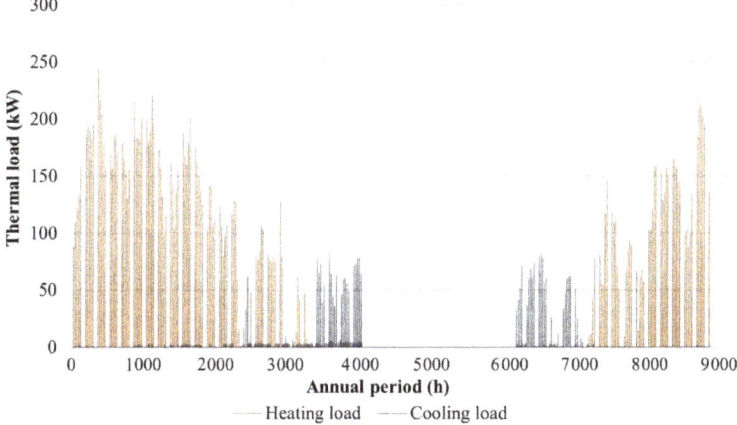

Figure 7. Annual time series of the heating and cooling loads of the examined school building.

In Figure 8a,b the heating loads fluctuation is depicted for the first two weeks of January and December respectively. In these figures it is seen that the heating loads are maximized with the beginning of the new working day, and then they gradually drop.

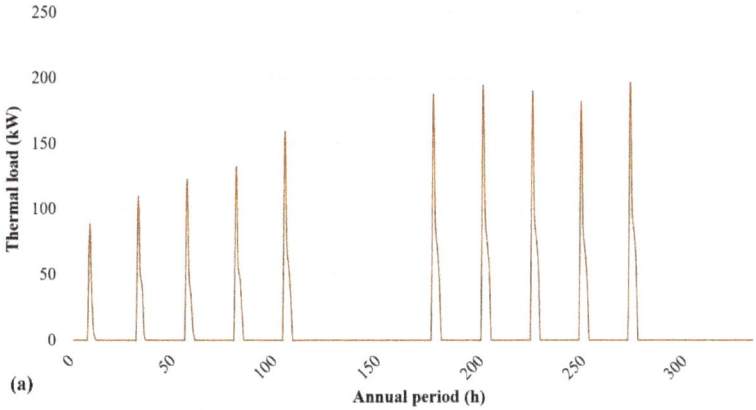

Figure 8. *Cont.*

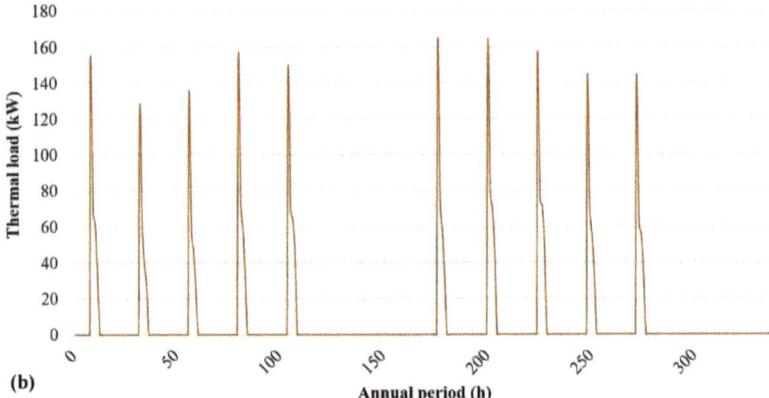

Figure 8. Heating loads fluctuation for the first two weeks of (**a**) January and (**b**) December.

This means that, practically, during the nights or the weekends, any potentially positive effect of the building's thermal inertia is lost. Additionally, all the uncovered heating loads during the inoperative periods are cumulated for the beginning of the next working day.

3. The Introduced Solar-Combi System

3.1. Scope

Within the framework of the energy upgrade of the examined school building, a solar-combi system was proposed and studied aiming at the 100% coverage of the indoor space heating loads. The solar-combi system was selected given the following very specific reasons:

- The high available solar radiation in the under consideration area and the availability of potentially abundant biomass fuel, mainly coming from the olive trees pruning.
- Greece holds the third place in Europe regarding the installation of solar collectors per capita, so far mainly for hot water production. This achievement has already enabled the development of a considerable domestic industry on the manufacturing of solar collectors. Additionally, the exploitation of the annually available huge amounts of pruning coming from the olive trees for the production of local biomass pellets can trigger the development of another significant local industry sector, contributing to the enforcement of the local economy.
- The installation of a solar-combi system in a school building can act as a pilot project, fostering the transition from the oil-based heating systems to the solar heating systems.
- This system can guarantee 100% coverage of the building's heating needs with locally available Renewable Energy Sources (solar radiation and biomass), substituting the currently imported oil and contributing, thus, further to the support of the local economy.

A new operation algorithm is developed for the examined solar-combi system. The objective of the introduced operation algorithm is the realization of the above scope (100% coverage of the indoor space heating loads) by maximizing the thermal power production by the solar collectors, reducing respectively the biomass consumption, while, at the same time, obtaining the minimum possible thermal energy annually levelized production cost. This cost will be approached with the following relationship:

$$\text{L.C.} = \frac{\frac{\text{I.C.}}{N} + \frac{\sum_{n=1}^{N} \frac{\text{A.O.C.}}{(1+i)^n}}{N}}{E_{th}} \qquad (1)$$

where:

L.C.: the annually average, thermal energy production levelized cost (in €/kWh$_{th}$)
I.C.: the initial cost (set-up cost) of the solar-combi system (in €)
A.O.C.: the total annual operation and maintenance cost (in €/year)
i: the discount rate, assumed equal to 3%
N: the total life period of the solar-combi system, assumed equal to 20 years
n: the number of the current year of the system's operation
E$_{th}$: the annual thermal energy production of the solar-combi system (kWh$_{th}$).

Given the request for 100% annual coverage of the building's indoor space heating, the annual thermal energy production E$_{th}$ is equal to the annual thermal energy demand, calculated in Table 3 equal to 32,356 kWh$_{th}$.

3.2. Operation Algorithm—Realization

The general layout of the solar-combi system is presented in Figure 9. It consists of the solar collectors' field, two, three or four thermal storage water tanks of 5000 lt volume capacity each and a back-up biomass heater. The number of the thermal storage tanks will be determined by the optimized dimensioning, which will be executed on the basis of the minimization of the levelized cost.

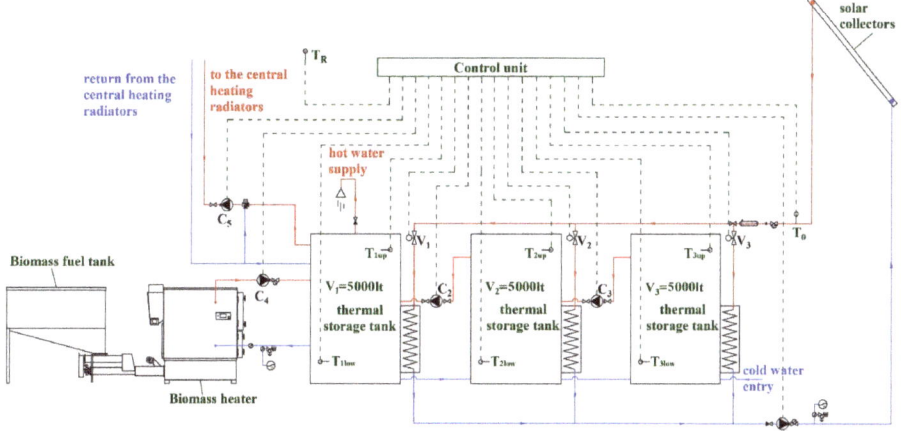

Figure 9. Graphical representation of the connectivity layout and the realization of the operation algorithm of the proposed system.

The biomass heater will be compatible with the relevant national and international standards (e.g., the ELOT 303-5 Greek national standard). The burner should be able to burn any kind of solid biomass fuel, such as pellets, olive kernels or any kind of small size solid biomass (e.g., nutshells). The overall efficiency of the biomass heater should be at the range of 85%. To facilitate the automatic operation of the whole system, the biomass heater should be equipped with all the technical specifications provided by the state of the art level of the relevant technology. Specifically:

- The biomass fuel will be automatically fed from the indoor fuel tank into the burner through a duct, operated by a conveyor, run by an electric motor.
- The burner will be equipped with an automatic ignition system with a blower, managed directly by the central control unit of the solar-combi system.
- The removal of the combustion residue should be facilitated with a specially designed system. Specifically, the residue will be collected to a removable bottom drawer. By removing this drawer, all the collected residue will be removed.

- To minimize any potential environmental impacts, the burner's chimney will be equipped with a cyclonic filter.

The type and the size of the introduced thermal tanks will support stratification thermal energy storage. This means that thermal energy will be stored inside the water thermal tanks maintaining a gradual increase of the water temperature from the tanks' bottom to the top, as illustrated in Figure 10.

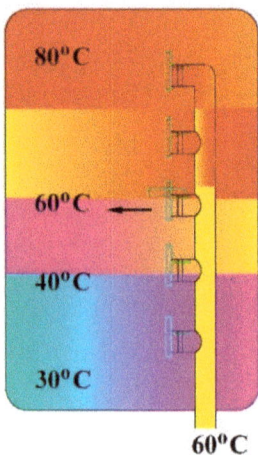

Figure 10. Stratification thermal energy storage in a water tank.

Stratification thermal storage is automatically achieved, through the transition of the warmer and lighter masses of the storage medium towards the upper layers of the thermal tank with physical flow, namely without any mechanical support. The required thermal power is always provided for the demand from the upper layer of the thermal tank, in order to ensure that thermal power will be distributed with the maximum achieved temperature. At the same time, the available for storage thermal power is provided for the thermal tank through a heat exchanger located close to the tank's bottom, in order to maximize the probability that the working fluid's temperature in the solar collectors' primary loop will be higher than the lowest temperatures existing in the thermal tank's low layers. In this way, the final storage of the initially available thermal power from the solar collectors' field is maximized.

The system's operation is automatically managed by a central control unit and facilitated through a number of thermometers, motor-vanes and circulators. Specifically, as seen in Figure 9, the control unit:

- receives signals from the thermometers T_R, T_0, T_{1low}, T_{1up}, T_{2low}, T_{2up}, T_{3low}, T_{3up}, from the lower and the upper layers of the thermal storage tanks, where T_R the temperature signal of the indoor space thermostat (or thermostats, in case more than one thermal zones of conditioned space with different required thermal comfort conditions are introduced)
- sends orders to the motor-vanes V_1, V_2 and V_3 and to the circulators C_1, C_2, C_3, C_4 and C_5.

The operation concept of the solar-combi system focuses on:

- the maintenance of the temperature inside the thermal tank 1 as high as possible, in order to minimize the biomass heater operation
- the maximization of the thermal energy storage from the solar collectors, even in cases this is available in relatively low temperatures.

The above objectives are simply approached by maintaining the highest water temperatures in thermal tank 1 and gradually decreasing temperatures in thermal tanks 2 and 3. In other words, the storage temperature in the available thermal storage tanks should be gradually kept lower from the tank 1 to the tank 3, in order to ensure that there will always be a storage tank with the minimum possible storage temperature, to maximize solar energy exploitation even in the early morning hours, when the working medium's temperature in the primary solar collectors loop is still relatively low. At the same time, the conservation of the maximum possible temperature in the thermal tank 1 will enable the minimization of the biomass heater operation. This storage tank will be the "load tank", since the central heating distribution network will be supplied exclusively from this tank. Additionally, as shown in Figure 9, the biomass heater will be also exclusively connected to this load tank.

The automatic operation of the combi-solar system under the above mentioned requirements is realized with the central control unit, the five circulating pumps, the three motorized valves, the eight temperature sensors installed as depicted in Figure 9 and the overall system's layout presented in the same figure. The above tasks are automatically realized with orders and the corresponding different operation modes presented in the lines below:

- if $T_R < T_{TC}$ then $C_5 = ON$
- if $T_0 > T_{1low}$ then: $C_1 = ON$, $V_1 =$ open, $V_2 =$ close, $V_3 =$ close
- if $T_0 < T_{1low}$ and $T_0 > T_{2low}$ then: $C_1 = ON$, $V_1 =$ close, $V_2 =$ open, $V_3 =$ close
- if $T_0 < T_{1low}$ and $T_0 < T_{2low}$ and $T_0 > T_{3low}$ then: $C_1 = OFF$, $V_1 =$ close, $V_2 =$ close, $V_3 =$ open
- if $T_{2up} > T_{1low}$ then: $C_2 = ON$
- if $T_{3up} > T_{2low}$ then: $C_3 = ON$
- if $T_{1up} < 70\ °C$ then: $C_4 = ON$.

where $T_{TC} = 22\ °C$ (see Section 2.3) the required thermal conditions temperature of the indoor conditioned space. The operation temperature range of the heating distribution network was set at 70–55 °C temperature. The supply temperature for the heating distribution network was set at this relatively low value to foster the penetration of the thermal power penetration in the system from the solar collectors and approach higher achieved performance efficiencies of the solar collectors.

3.3. Simulation Methodology

The dimensioning procedure was based on the arithmetic simulation of the annual operation of the combi-solar system, based on the fundamental theory of solar collectors and thermal energy storage. The full mathematical background of the executed simulation process is analytically presented in [44]. The solar collectors' performance was simulated with the TRNSYS software application and the annual thermal power production time series was developed respectively. The employed TRNSYS model is presented in Figure 11. The simulation was executed for 5 different scenarios regarding the size of the solar collectors' field, with 36, 40, 44, 48 and 52 solar collectors involved in each one of them, with an effective collector's surface of 2.30 m². The solar collectors for each different scenarios were divided in parallel groups of 4 collectors connected in series, as presented in Figure 12.

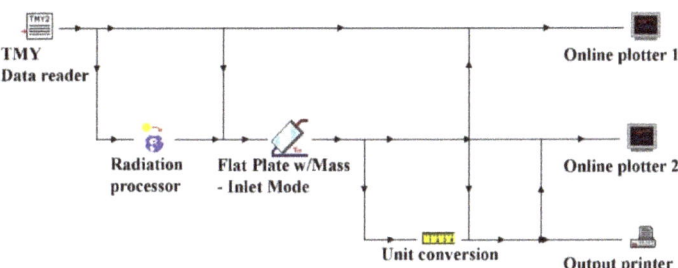

Figure 11. Calculation logical diagram introduced in TRNSYS for the simulation of the solar collectors operation.

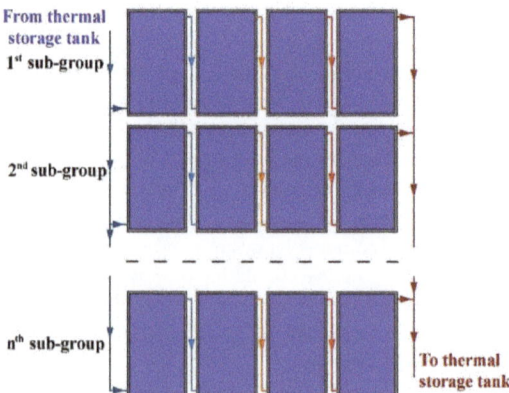

Figure 12. Connection layout of the solar collectors' field.

The TRNSYS simulation model was also utilized for the determination of the optimum installation angle of the solar collectors versus the horizontal plane. This was executed aiming at the maximisation of the thermal energy production during the winter period (from the 1st of November to the 31st of March) and for the geographical latitude of the under consideration location (35°). By executing similar simulations of the solar collectors' performance under alternative installation angles with an increment step of 5°, it was concluded that the optimum installation angle is 45°. Obviously the solar collectors are oriented to the south (collectors' surface azimuth angle equal to 0°). For this optimum installation angle and orientation, the annual variation of the total incident solar radiation (direct, diffused and reflected) per unit of solar collectors' surface (in W/m^2) is depicted in Figure 13. The simulation process is based on the following essential steps:

- The simulation is executed on hourly, average calculation time steps.
- For every hourly calculation step, the thermal power production P_{sc} from the solar collectors' field is provided from the corresponding time series developed with TRNSYS.
- If P_{td} is the thermal power demand, then the direct thermal power production penetration P_{sp} from the solar collectors' field to the thermal power demand coverage is simply calculated as:

if $P_{td} \geq P_{sc}$ then $P_{sp} = P_{sc}$;
if $P_{td} < P_{sc}$ then $P_{sp} = P_{td}$.

- The thermal power storage P_{sta} from the solar collectors will in any case be:

$P_{sta} = P_{sc} - P_{sp}$.

- The remaining thermal power demand P_{tdr}, after the direst penetration from the solar collectors will be:

$P_{tdr} = P_{td} - P_{sp}$.

- The remaining thermal energy demand and the total thermal energy storage from the solar collectors are calculated for every 24-h period, by integrating the corresponding 24-h thermal power time series.
- For the current 24-h period, the thermal power demand coverage from the stored thermal energy from the previous 24-h period is calculated. Any possible remaining stored power will be utilized for the next 24-h period (this mainly happens during the late autumn or the early spring period).
- Finally, any remaining thermal power demand, after the exploitation of the stored thermal power will be covered by the biomass heater. This thermal energy is also calculated on a 24-h basis.

- Given the above approach, the dimensioning of the required thermal storage tank is imposed by the maximum required thermal energy storage from the solar collectors for a 24-h period during the year.
- For each different dimensioning scenario, regarding the required thermal storage capacity and the solar collectors total number, the annual time series of mean hourly or daily values are calculated for the:
    - initial thermal power production from the total solar collectors' field
    - the solar collectors' direct thermal power penetration for the indoor space heating loads coverage
    - the thermal power storage from the solar collectors
    - the thermal power production from the biomass heater.
- Annual statistics for the thermal energy produced and stored from the solar collectors and the biomass heater are eventually calculated by integrating the above mentioned developed corresponding annual time series.

The results from the execution of the applied methodology are presented in the next section.

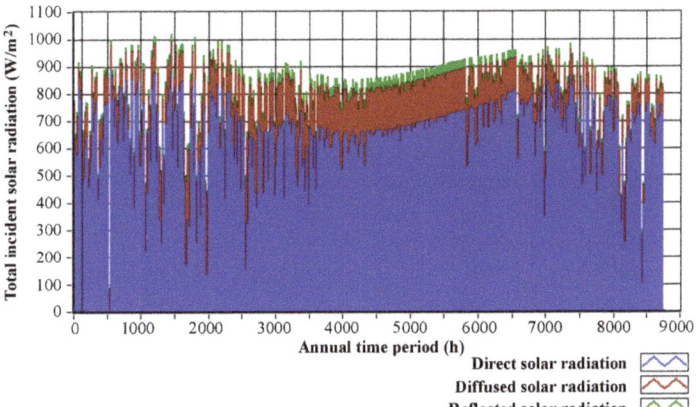

Figure 13. Annual variation of the total incident solar radiation for the installation angle of 45° and for south orientation.

3.4. Results

The essential results from the annual simulation of the solar-combi system's operation regarding the annual energy production and storage are presented in Table 4.

Table 4. Annual thermal energy production and storage results.

Number of Solar Collectors	36	40	44	48	52
Solar collectors total surface (m^2)	82.8	92	101.2	110.4	119.6
Solar collectors production from 15/10—15/4	17,072	18,774	20,480	22,097	23,801
Solar collectors direct thermal energy penetration	4711	5093	5448	5803	6157
Solar collectors thermal energy available for storage	12,361	13,680	15,031	16,294	17,644
Solar collectors thermal energy eventually stored	7677	8354	9068	9721	10,292
Biomass heat thermal energy production	19,968	18,909	17,840	16,833	15,907
Total thermal energy production	32,356	32,356	32,356	32,356	32,356
Required thermal storage capacity (kWh$_{th}$)	198.98	220.25	241.81	262.76	283.82
Required water tanks capacity (kg)	11,370	12,586	13,818	15,015	16,219
Solar collectors' annual percentage coverage (%)	38.29	41.56	44.86	47.98	50.84
Solar collectors' thermal energy annual percentage surplus (%)	27.43	28.37	29.12	29.75	30.89

By observing the results presented in Table 4, it is possible to proceed to the following remarks:

- The annual contribution of the solar collectors to the thermal energy demand coverage ranges from 38% to 50%, for the different investigating scenarios, versus the total installed solar collectors' surface. The remaining thermal energy demand is covered with the biomass heater.
- Percentage 27–28% of the total thermal energy contribution from the solar collectors comes from direct penetration and the rest percentage comes from the utilization of the stored thermal energy.
- The annual thermal energy surplus from the solar collectors is relatively lowly restricted, namely from 27% to 30%. This verifies the appropriate dimensioning of the required thermal storage tanks. On the other hand, this low annual thermal energy surplus is sensible and should be expected, given the fact that:

    - the simulation refers to the winter period, during which the available solar radiation is relatively low
    - at the same period, there is a considerable heating load of a school building with covered area of indoor conditioned spaces higher than 1000 m², supposed to be covered by this particular solar collectors' field.

In Table 5, the results from the calculation of the thermal energy production levelized cost are presented. The following set-up costs and procurement prices have been adopted, based on commercial quotations from equipment manufacturers and suppliers:

- biomass heater and accessories procurement—installation cost: 30,150 €
- solar collector with selective coating procurement cost: 220 €
- water thermal storage tank of 5000 lt capacity procurement cost: 10,000 €
- remaining hydraulic and electronic equipment procurement and installation cost: 8000 €
- biomass pellets procurement price (in Crete): 350 €/tn
- the annual average maintenance and operation cost is configured by the consumed biomass pellets procurement cost and the biomass heater annual maintenance cost, set, on average, equal to 200 €.

Table 5. Calculation of the thermal energy production annual average levelized cost.

Cost Component	Investigating Scenario (Number of Solar Collectors/Thermal Tanks)					
	36/3	40/3	44/3	48/3	52/4	52/3
Biomass heater cost (€)	30,150	30,150	30,150	30,150	30,150	30,150
Solar collectors cost (€)	7920	8800	9680	10,560	11,440	11,440
Thermal storage tanks cost (€)	30,000	30,000	30,000	30,000	40,000	30,000
Rest equipment cost (€)	8000	8000	8000	8000	8000	8000
Total set-up cost (€)	76,070	76,950	77,830	78,710	89,590	79,590
Biomass pellets annual consumption (tn)	3.840	3.636	3.431	3.237	3.059	3.096
Average annual maintenance and operation cost (€)	1200	1147	1093	1043	996	1006
Thermal energy production levelized cost (€/kWh$_{th}$)	0.1546	0.1544	0.1541	0.1539	0.1692	0.1541

Finally, the lowest calorific value of the biomass pellets was assumed equal to 5.2 kWh/kg. From the results presented in Table 5, it is concluded that the optimum size for the introduced solar-combi system, with regard to the minimization of the thermal energy production levelized cost, is configured with the installation of 48 solar collector with selective coating and 2.3 m² of effective surface and three thermal storage tanks of 5000 L water storage capacity each. The minimum achieved levelized cost for thermal energy production, based on the above mentioned assumptions and prices, is calculated at 0.1539 €/kWh$_{th}$. For the next investigating scenario, with the installation of 52 solar collectors, one additional thermal storage tank is required, raising considerably the total set-up cost and the corresponding production levelized cost. Nevertheless, it should be noticed that even if the number of the thermal storage tanks is theoretically kept at 3, instead of 4, for the dimensioning scenario of 52 solar collectors, the arisen production levelized cost is calculated at 0.1541 €/kWh$_{th}$, namely higher than the corresponding figure of the scenario with 48 collectors, which still remains the optimum one.

In Figure 14, power production synthesis curves are presented for (a) the period from the 1st of January to the 15th of April and (b) the period from the 15th of October to the 31st of December. Interesting conclusions are derived from these figures:

- After the weekends, the stored thermal energy in the thermal storage tanks undertakes most of the thermal power demand for the first days of the new week.
- While approaching the last days of March or during the last days of October, the thermal power direct penetration of the solar collectors and the contribution of the thermal storage tanks undertakes all the heating loads. The biomass heater contribution during these periods is negligible.

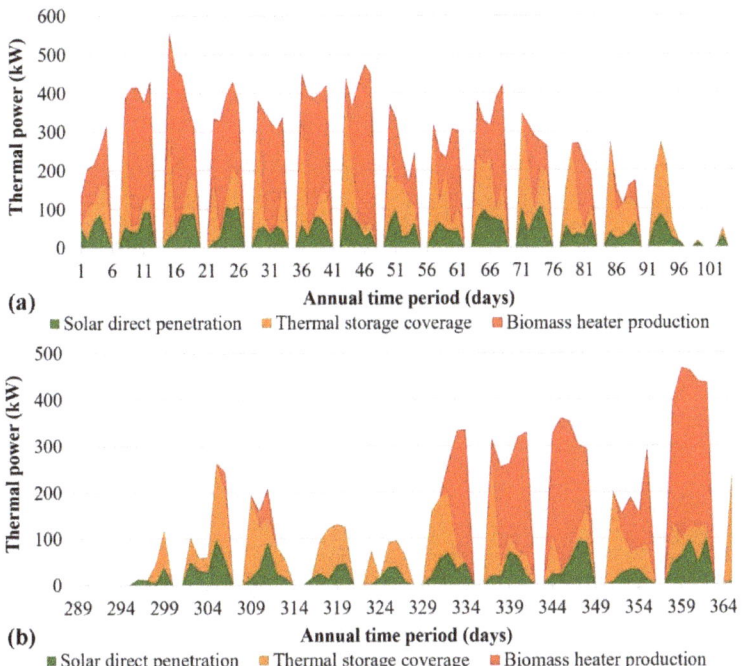

Figure 14. Thermal power production synthesis from the solar-combi system components for (**a**) the period from 1/1 to 15/4 and (**b**) the period from 15/10 to 31/12.

Given the annual operation and maintenance cost of the existing oil-based central heating system (5080 €, see Section 2.3) and the corresponding figure of the optimized proposed solar-combi system (1043 €, Table 5) the net annual economic benefit from the solar-combi system introduction and the elimination of the diesel oil consumption equals to 4037 €. Given also the total set-up cost of optimized solar-combi system (78,710 €, Table 5), its payback period is calculated at 19.50 years. This figure is certainly quite high, yet, the following facts should be taken into account:

- The specific school building has a limited operation period for roughly 8–10 h per day and for five days per week. Additionally, the existing thermal energy needs in the building refer only to the indoor space heating loads, appeared only for six months per year.

- No other thermal energy needs, apart from the indoor space heating, are present due to the lack of additional facilities, e.g., gyms accompanied with changing rooms and showers, swimming pools, dormitories etc.
- The economic feasibility of the same system in another type of building at the same geographical location (e.g., a residence or a sports hall with extensive hot water needs) is expected much higher. In this case, the annual average thermal energy production cost should be also expected much lower than the calculated figures in this article.
- However, the specific project has a strong demonstrative aspect, given the daily use of the school building from students, namely from people with their educational background still under cultivation. The proposed system, along with the rest energy saving measures proposed in the frame of the building's energy performance upgrade, aim to contribute, apart from the obvious energy saving target, to the cultivation of the rational energy use culture for the building's users and visitors.
- Finally, since the project was 100% funded by a national—E.U. co-funding call, the economic feasibility of the proposed system was not the crucial decision and design parameter.

3.5. Practical Issues—Drawbacks

The introduced solar-combi system, as shown above, is able to guarantee 100% annual heating loads coverage of the examined school building, with annual average heating production levelized cost lower than the corresponding feature of the existing oil-based central heating system. Yet, the proposed system exhibits some very specific practical issues, which, however, can be easily handled, as explained below:

- For six months per year, namely from mid-April to mid-October, the system will remain inoperative, due to the lack of any heating loads in the building, or any other thermal energy needs, as mentioned in the previous section. Due to this fact, in order to avoid over heating of the water in the thermal storage tanks, the solar collectors' transparent glazing should be covered with opaque textiles.
- A regular cleaning process for the solar collectors' glazing should be regularly applied, indicatively once a year, most probably at the beginning of the heating season.
- A regular cleaning of the biomass heater (removal of the combustion residue) should be also regularly performed, indicatively once a week. This is, objectively, the most demanding required task.

Finally, an appropriate infrastructure should be constructed, consisting of a large, outdoor biomass fuel closed tank and a duct, for the automatic refilling of a second, smaller silo inside the machinery room, from which the biomass heater will be supplied.

4. Conclusions

A solar-combi system was investigated in this article for the indoor space heating of a school building, located in the mainland of the island of Crete. The specific building exhibits two major favorable features, regarding the coverage of its heating loads from solar collectors:

- Firstly, there is considerable solar radiation available in the particular geographical location even during winter, a crucial issue for the maximisation of the building's heating loads coverage from solar collectors.
- Secondly, another important parameter, perhaps not so obvious, is the intermittent operation of the school building. This operational feature has a considerable contribution to the achievement of high coverage percentages of the building's heating loads from the solar collectors. Specifically, the lack of heating loads after the end of the school's daily operation schedule, on the one hand contributes to the reduction of the building's heating loads (no heating loads are considered during non-operational time periods), while, on the other hand, given the appropriate sizing of the thermal storage tanks, the thermal energy stored during non-operational periods, especially

during the weekends, is utilized for the coverage of the forthcoming heating loads. In this way, the potential for high solar collectors' penetration to the annual thermal energy demand coverage is remarkably increased.

The two favorable parameters mentioned above enable the achievement of annual coverage percentages of the introduced solar collectors' field in the examined school building higher than 50%, of course given the effective introduced algorithm and its adequate realization with the proposed system's overall layout.

On the other hand, the exclusive thermal energy consumption for the indoor space heating and the lack of any other thermal energy needs, has a negative impact on the economic feasibility of the examined system. The long payback period calculated at 19.50 years should be expected much shorter in case of a building with additional thermal energy needs (e.g., hot water consumption or longer operation period per day etc.).

The successful simulation and dimensioning of the examined system is also confirmed by the low surplus of the thermal energy produced from the solar collectors. This result should be absolutely expected too, given the large size of the building with the corresponding heating load and the, although still remarkable, reduced, compared to the summer period, available solar radiation. Nevertheless, all the favorable conditions mentioned above seem to be effectively exploited in the synthesis of the proposed system and the operation concept indicated with the introduced algorithm. The last step of course is the realization of the proposed system in the under consideration school building and the confirmation in practice of the simulation results of the present article.

In any case, the present article demonstrates a specific layout and operation algorithm for a solar-combi system, as well as the way this system can be realized and its automatic operation can be performed. Beyond any technical aspect, it certainly demonstrates the feasibility of 100% heating loads coverage in buildings, based on R.E.S. (solar radiation and biomass), especially in Southern European climates. This can be achieved with absolutely competitive production levelized cost, which, in the case of the examined case study, ranges at 0.154 €/kWh$_{th}$. This cost is lower than the corresponding figure with the use of a diesel oil central heater for the heating of the same school building, calculated at 0.1570 €/kWh$_{th}$.

Author Contributions: Conceptualization, D.A.K.; methodology, D.A.K. and G.Z.; software, D.A.K. and G.Z.; validation, D.A.K. and G.Z.; formal analysis, D.A.K. and G.Z.; investigation, D.A.K.; resources, D.A.K. and G.Z.; data curation, G.Z.; writing—original draft preparation, D.A.K.; writing—review and editing, D.A.K.; visualization, D.A.K.; supervision, D.A.K.; project administration, D.A.K.

Funding: This research received no external funding.

Acknowledgments: The authors of this study gratefully acknowledge the assistance and the contribution of Gareth Owens, for his contribution to the linguistic integrity of the article.

Conflicts of Interest: The authors declare no conflict of interest.

References

1. U.S. Energy Information Administration. 2012 Commercial Buildings Energy Consumption Survey: Energy Usage Summary. Available online: https://www.eia.gov/consumption/commercial/reports/2012/energyusage/ (accessed on 1 November 2018).
2. Behar, O. Solar thermal power plants—A review of configurations and performance comparison. *Renew. Sustain. Energy Rev.* **2018**, *92*, 608–627. [CrossRef]
3. Gorjian, S.; Ghobadian, B. Solar thermal power plants: Progress and prospects in Iran. *Energy Procedia* **2015**, *75*, 533–539. [CrossRef]
4. Cunio, L.N.; Sproul, A.B. Performance characterisation and energy savings of uncovered swimming pool solar collectors under reduced flow rate conditions. *Sol. Energy* **2012**, *86*, 1511–1517. [CrossRef]
5. Bunea, M.; Perers, B.; Eicher, S.; Hildbrand, C.; Bony, J.; Citherlet, S. Mathematical modelling of unglazed solar collectors under extreme operating conditions. *Sol. Energy* **2015**, *118*, 547–561. [CrossRef]

6. Soltau, H. Testing the thermal performance of uncovered solar collectors. *Sol. Energy* **1992**, *49*, 263–272. [CrossRef]
7. Hashim, W.M.; Shomran, A.T.; Jurmut, H.A.; Gaaz, T.S.; Al-Amiery, A.A. Case study on solar water heating for flat plate collector. *Case Stud. Therm. Eng.* **2018**, *12*, 666–671. [CrossRef]
8. Pandey, K.M.; Chaurasiya, R. A review on analysis and development of solar flat plate collector. *Renew. Sustain. Energy Rev.* **2017**, *67*, 641–650. [CrossRef]
9. Hao, W.; Lu, Y.; Lai, Y.; Yu, H.; Lyu, M. Research on operation strategy and performance prediction of flat plate solar collector with dual-function for drying agricultural products. *Renew. Energy* **2018**, *127*, 685–696. [CrossRef]
10. Tian, Z.; Perers, B.; Furbo, S.; Fan, J. Thermo-economic optimization of a hybrid solar district heating plant with flat plate collectors and parabolic trough collectors in series. *Energy Convers. Manag.* **2018**, *165*, 92–101. [CrossRef]
11. Nitsas, M.T.; Koronaki, I.P. Experimental and theoretical performance evaluation of evacuated tube collectors under Mediterranean climate conditions. *Therm. Sci. Eng. Prog.* **2018**, *8*, 457–469. [CrossRef]
12. Shafieian, A.; Khiadani, M.; Nosrati, A. A review of latest developments, progress, and applications of heat pipe solar collectors. *Renew. Sustain. Energy Rev.* **2018**, *95*, 273–304. [CrossRef]
13. Daghigh, R.; Shafieian, A. Theoretical and experimental analysis of thermal performance of a solar water heating system with evacuated tube heat pipe collector. *Appl. Therm. Eng.* **2016**, *103*, 1219–1227. [CrossRef]
14. Sabiha, M.A.; Saidur, R.; Mekhilef, S.; Mahian, O. Progress and latest developments of evacuated tube solar collectors. *Renew. Sustain. Energy Rev.* **2015**, *51*, 1038–1054. [CrossRef]
15. Lamnatou, C.; Cristofari, C.; Chemisana, D.; Canaletti, J.L. Building-integrated solar thermal systems based on vacuum-tube technology: Critical factors focusing on life-cycle environmental profile. *Renew. Sustain. Energy Rev.* **2016**, *65*, 1199–1215. [CrossRef]
16. Hassanien, R.; Hassanien, E.; Li, M.; Tang, Y. The evacuated tube solar collector assisted heat pump for heating greenhouses. *Energy Build.* **2018**, *169*, 305–318. [CrossRef]
17. Shafii, M.B.; Mamouri, S.J.; Lotfi, M.M.; Mosleh, H.J. A modified solar desalination system using evacuated tube collector. *Desalination* **2016**, *396*, 30–38. [CrossRef]
18. Tyagi, V.V.; Kaushik, S.C.; Tyagi, S.K. Advancement in solar photovoltaic/thermal (PV/T) hybrid collector technology. *Renew. Sustain. Energy Rev.* **2012**, *16*, 1383–1398. [CrossRef]
19. Abdelrazik, A.S.; Al-Sulaiman, F.A.; Saidur, R.; Ben-Mansour, R. A review on recent development for the design and packaging of hybrid photovoltaic/thermal (PV/T) solar systems. *Renew. Sustain. Energy Rev.* **2018**, *95*, 110–129. [CrossRef]
20. Parvez Mahmud, M.A.; Huda, N.; Farjana, S.H.; Lang, C. Environmental impacts of solar-photovoltaic and solar-thermal systems with life-cycle assessment. *Energies* **2018**, *11*, 2346. [CrossRef]
21. Kefalloniti, I.; Ampatzi, E. Building integration of domestic solar combi-systems: The importance of managing the distribution pipework. *Energy Build.* **2017**, *142*, 179–190. [CrossRef]
22. Glembin, J.; Haselhorst, T.; Steinweg, J.; Rockendorf, G. Simulation and evaluation of solar thermal combi systems with direct integration of solar heat into the space heating loop. *Energy Procedia* **2016**, *91*, 450–459. [CrossRef]
23. Antoniadis, C.N.; Martinopoulos, G. Simulation of solar thermal systems with seasonal storage operation for residential scale applications. *Procedia Environ. Sci.* **2017**, *38*, 405–412. [CrossRef]
24. Yang, Z.; Wang, Y.; Li, Z. Building space heating with a solar-assisted heat pump using roof-integrated solar collectors. *Energies* **2011**, *4*, 504–516. [CrossRef]
25. Kalder, J.; Annuk, A.; Allik, A.; Kokin, E. Increasing solar energy usage for dwelling heating, using solar collectors and medium sized vacuum insulated storage tank. *Energies* **2018**, *11*, 1832. [CrossRef]
26. Katsaprakakis, D.A.; Zidianakis, G. Upgrading energy efficiency for school buildings in Greece. *Procedia Environ. Sci.* **2017**, *38*, 248–255. [CrossRef]
27. Kyriaki, E.; Giama, E.; Papadopoulou, A.; Drosou, V.; Papadopoulos, A.M. Energy and environmental performance of solar thermal systems in hotel buildings. *Procedia Environ. Sci.* **2017**, *38*, 36–43. [CrossRef]
28. Calise, F.; Figaj, R.D.; Vanoli, L. Energy and economic analysis of energy savings measures in a swimming pool centre by means of dynamic simulations. *Energies* **2018**, *11*, 2182. [CrossRef]
29. Katsaprakakis, D.A. Comparison of swimming pools alternative passive and active heating systems based on renewable energy sources in Southern Europe. *Energy* **2015**, *81*, 738–753. [CrossRef]

30. Hsieh, S.; Omu, A.; Orehounig, K. Comparison of solar thermal systems with storage: From building to neighbourhood scale. *Energy Build.* **2017**, *152*, 359–372. [CrossRef]
31. Tirumala, N.; Kumar, U.; Martin, A.R. Co-production performance evaluation of a novel solar combi system for simultaneous pure water and hot water supply in urban households of UAE. *Energies* **2017**, *10*, 481. [CrossRef]
32. Bois, J.; Mora, L.; Wurtz, E. Energy saving analysis of a solar combi-system using detailed control algorithm modeled with Modelica. *Energy Procedia* **2015**, *78*, 1985–1990. [CrossRef]
33. Deng, S.; Dai, Y.J.; Wang, R.Z. Performance optimization and analysis of solar combi-system with carbon dioxide heat pump. *Sol. Energy* **2013**, *98*, 212–225. [CrossRef]
34. D'Antoni, M.; Ferruzzi, G.; Bettoni, D.; Fedrizzi, R. Validation of the numerical model of a turnkey solar combi + system. *Energy Procedia* **2012**, *30*, 551–561. [CrossRef]
35. Johansen, J.B.; Englmair, G.; Dannemand, M.; Kong, W.; Furbo, S. Laboratory testing of solar combi system with compact long term PCM heat storage. *Energy Procedia* **2016**, *91*, 330–337. [CrossRef]
36. Englmair, G.; Dannemand, M.; Johansen, J.B.; Kong, W.; Dragsted, J.; Furbo, S.; Fan, J. Testing of PCM heat storage modules with solar collectors as heat source. *Energy Procedia* **2016**, *91*, 138–144. [CrossRef]
37. Colclough, S.; McGrath, T. Net energy analysis of a solar combi system with Seasonal Thermal Energy Store. *Appl. Energy* **2015**, *147*, 611–616. [CrossRef]
38. Ma, Z.; Bao, H.; Roskilly, A.P. Feasibility study of seasonal solar thermal energy storage in domestic dwellings in the UK. *Sol. Energy* **2018**, *162*, 489–499. [CrossRef]
39. Andersen, E.; Furbo, S. Theoretical variations of the thermal performance of different solar collectors and solar combi systems as function of the varying yearly weather conditions in Denmark. *Sol. Energy* **2009**, *83*, 552–565. [CrossRef]
40. Hellenic National Meteorological Service: Meteorological Stations in Crete: Kasteli Pediados. Available online: http://hnms.gr/emy/el/observation/sa_teleytaies_paratiriseis_stathmou?perifereia=Crete&poli=Kasteli_Pediados (accessed on 22 November 2018).
41. Kreider, J.; Rabl, A.; Curtiss, P. *Heating and Cooling of Buildings*, 3rd ed.; CRC Press: Boca Raton, FL, USA, 2017.
42. Official Governmental Gazette 2367B'/12-7-2017. Directive on Buildings' Energy Performance. Available online: http://tdm.tee.gr/wp-content/uploads/2017/07/fek_12_7_2017_egrisi_kenak.pdf (accessed on 4 January 2019).
43. *2009 ASHRAE Handbook—Fundamentals (SI Edition)*; American Society of Heating, Refrigerating and Air-Conditioning Engineers, Inc.: Atlanta, GA, USA, 2009.
44. Duffie, J.A.; Beckman, W.A. *Solar Engineering of Thermal Processes*, 4th ed.; Wiley: New York, NY, USA, 2013.

© 2019 by the authors. Licensee MDPI, Basel, Switzerland. This article is an open access article distributed under the terms and conditions of the Creative Commons Attribution (CC BY) license (http://creativecommons.org/licenses/by/4.0/).

Article

Non-Iterative Methods for the Extraction of the Single-Diode Model Parameters of Photovoltaic Modules: A Review and Comparative Assessment

Efstratios Batzelis

Department of Electrical and Electronic Engineering, Imperial College London, London SW7 2AZ, UK; e.batzelis@imperial.ac.uk

Received: 10 December 2018; Accepted: 19 January 2019; Published: 23 January 2019

Abstract: The extraction of the photovoltaic (PV) model parameters remains to this day a long-standing and popular research topic. Numerous methods are available in the literature, widely differing in accuracy, complexity, applicability, and their very nature. This paper focuses on the class of non-iterative parameter extraction methods and is limited to the single-diode PV model. These approaches consist of a few straightforward calculation steps that do not involve iterations; they are generally simple and easy to implement but exhibit moderate accuracy. Seventeen such methods are reviewed, implemented, and evaluated on a dataset of more than one million measured I-V curves of six different PV technologies provided by the National Renewable Energy Laboratories (NREL). A comprehensive comparative assessment takes place to evaluate these alternatives in terms of accuracy, robustness, calculation cost, and applicability to different PV technologies. For the first time, the irregularities found in the extracted parameters (negative or complex values) and the execution failures of these methods are recorded and are used as an assessment criterion. This comprehensive and up-to-date literature review will serve as a useful tool for researchers and engineers in selecting the appropriate parameter extraction method for their application.

Keywords: analytical; explicit; five parameters; Lambert W function; parameters extraction; photovoltaic (PV); review; single-diode model

1. Introduction

The model of a photovoltaic (PV) generator usually consists of an equivalent circuit and a set of parameters that describe its electrical response and operation. Determination of these parameters is not a trivial task, as they are not available in the PV module's datasheet and their values change with the operating conditions. Immense research has been carried out in recent decades on the extraction of the PV model parameters, the literature presenting numerous methods of different nature, reliability, complexity, and required input data. It now constitutes a research topic on its own, referred to as "PV cell model parameters estimation problem" or similar in the literature [1].

These methods can be classified into three major categories: the *numerical*, the *non-iterative* and the *optimization* approaches. The numerical (or *iterative*) methods form a system of a few equations which is solved numerically [2–6], in a trial-and-error manner or via another iterative algorithm [7–10]. These equations are usually derived by applying the PV model equation to specific conditions, such as short-circuit (SC), open-circuit (OC) or maximum power point (MPP). This class achieves generally high accuracy but suffers from initialization and convergence issues, high calculation cost and solution suboptimality [1,11–13]. The non-iterative (or *explicit* or *direct* or *analytical*) methods employ a set of equations as well, but are solved symbolically/explicitly (no iterations) resulting in simpler formulation and implementation [12,14–28]. These approaches are essentially variations of the numerical class that employ a series of simplifications and empirical observations to achieve explicit formulation. Quite often, they are used in the initialization step of the numerical methods [26,29]. It is worth noting that the term "analytical" is somehow ambiguous in the literature, referring either to the non-iterative, numerical, or both classes, thus it is not used in this paper to avoid confusion. The non-iterative methods are easier to implement and more computationally efficient but yield lower accuracy [13], although some of these approaches perform quite decently [30]. The optimization (or *artificial intelligence* or *heuristic* or *curve-fitting* or *soft-computing*) methods follow a non-technologically specific approach where the model's equation is optimally fitted on a set of measurements, usually the *I-V* curve. Various evolution inspired [31–33] and curve-fitting [34,35] algorithms may be found in the literature. This class exhibits generally high accuracy and near-global optimality but suffers from computational complexity and difficulties in the method's parameters tuning [1].

Another important aspect of a parameter extraction technique is the required input data. For some methods, the datasheet information suffices (e.g., SC, OC, and MPP data, temperature coefficients etc.) [12,15,17,20–22,25–27], while others require additional operating points or/and the slope of the *I-V* curve at SC or OC [14,16,18,19,23,24]. Generally, the former cases are preferable as they can be applied more easily and universally, not necessitating extra measurements [19,27,36]. Furthermore, there is currently a debate in the literature on whether the extracted parameters should be restricted to real positive numbers to have a physical meaning [4,26], or should be allowed to take negative or complex values (referred to as *parameter irregularities* in this paper) if the resulted curve better matches the measurements [15,30]. Other desirable attributes of such a method is to be accurate, robust, computationally efficient, and applicable to various PV technologies [1,27].

The large diversity in the requirements, performance and very nature of these methods has inspired the publication of several review papers in the literature [1,30,36–41]. The studies in [1,37] provide a qualitative description and classification of all three classes, but without implementation and quantitative comparison. The rest of the aforementioned papers are limited to the numerical or/and non-iterative methods, implementing and evaluating some of them. However, only a small subset of the available non-iterative methods is assessed, and the evaluation dataset corresponds mainly to single- and multi-crystalline silicon PV modules [30,36,38]. The copper indium gallium selenide (CIGS) and cadmium telluride (CdTe) technologies are considered in [39], heterojunction with intrinsic thin layer (HIT) modules are tested in [40] and a few multi-junction devices are examined in [41]. To this day, there is no comparative study in the literature to consider all the commercially available PV technologies. Furthermore, although the robustness has been investigated in the numerical and optimization classes, the aspect of execution failure or irregular results in the non-iterative methods has not been studied yet.

To shed some light on these topics, this paper performs a comprehensive review and comparison of the non-iterative parameter extraction methods that are based on the *single-diode PV model*. Seventeen such methods are identified dating since 1984 [12,14–27], all of which are implemented and assessed on a common set of *I-V* curves provided by National Renewable Energy Laboratories (NREL) [42]. This dataset contains 1,025,665 *I-V* curves measured over a one-year period in the USA for 22 PV modules under a wide range of irradiance and temperature conditions. This is probably one of the most comprehensive publicly available datasets, including six different PV technologies: single- (c-Si) and multi-crystalline (mc-Si), CdTe, CIGS, HIT and amorphous silicon (a-Si) (crystalline, tandem, and triple-junction). It is worth noting that this investigation is limited to the single-diode PV model, adopted by most of the non-iterative parameter extraction methods, even though other more sophisticated models may be more appropriate for some thin-film technologies.

The comparison that follows aims to give the full picture on the appropriateness of the non-iterative methods in terms of accuracy, robustness, and calculation cost for all these PV technologies. Special focus is given on the *execution failures* recorded for the 1 million scenarios to evaluate their credibility, and the number of *parameter irregularities* (negative or complex values) that may be or may be not a limiting factor to some applications. The assessment is performed first for all 17 methods, and then separately for those that rely solely on datasheet information; furthermore, a sensitivity analysis takes place on the fitting range for the extraction of the SC slope. This survey is a thorough and up-to-date comparative assessment of the available non-iterative methods, carried out for the first time on a comprehensive dataset and accounting for the aspects of reliability and robustness.

The rest of the paper is organized as follows: the basics of the single-diode PV model are given in Section 2 and the non-iterative parameter extraction methods are described and discussed in Section 3. The performance of these methods is assessed and compared in Section 4, the main conclusions summarized in Section 5. The Appendixes A and B clarifies some calculation aspects.

2. The Fundamentals of the Single-Diode PV Model

This section gives a brief description of the PV model theory under consideration; the following nomenclature is used throughout the paper for consistency. The single-diode model is historically the first PV model, developed initially for single-crystalline silicon PV cells, but it remains to this day the most commonly used one due to its simplicity [2,43,44]. Other more sophisticated models involve two [45] or three diodes [46] for increased accuracy at low irradiance, and sometimes additional voltage-dependent current sources to account for the breakdown operation [47] or the recombination phenomenon in some thin-film technologies [48]. This paper is limited to the single-diode model, as this is the PV model on which the majority of the *non-iterative* parameter extraction methods are based, and assesses its effectiveness on all commercial PV technologies, single/multi-crystalline silicon and thin-film.

This model consists of an equivalent circuit shown in Figure 1 and a set of *five parameters* $[I_{ph}, I_s, a, R_s, R_{sh}]$:

- the photocurrent I_{ph} (or I_0 or I_{pv})
- the diode saturation current I_s (or I_0 or I_{sat})
- the modified diode factor a (or m)
- the series resistance R_s
- the shunt resistance R_{sh} (or R_p).

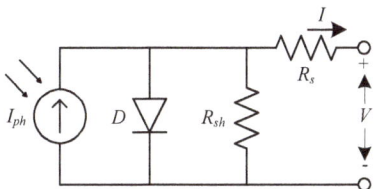

Figure 1. Electrical equivalent circuit of the single-diode PV model.

Quite often, the modified diode factor a is expressed alternatively as $N_s n V_T$ in the literature, where N_s is the number of series-connected cells, n the diode factor and $V_T = \frac{kT}{q}$ the thermal voltage (k is the Boltzmann constant, T the temperature in Kelvin and q the electron charge). The five parameters depend on both the structural characteristics of the PV modules and the operating conditions: incident irradiance and cell temperature. The irradiance is usually considered to affect proportionally I_{ph} [1–4,7,20,21,27,28,36,43,49–54], and R_{sh} in an inversely proportional way [1–4,20,27,36,43,49,51,52,54]; the temperature effect is generally assumed to be weak and linear in I_{ph}, strong and exponential in I_s, and strong and proportional in a [1–4,7,20,21,27,36,43,49–54]. The dependence of R_s is somewhat unclear, some studies assuming to remain constant [2–4,36,43,50,51,53,54] and others to depend on both irradiance and temperature [1,20,27,49,52]. Typically, the five parameters are not assumed to be affected by the operating point, i.e., they do not change along the *I-V* curve. For some technologies, however, it has been reported to vary with the current, especially the diode factor and series resistance; here, the former approach is considered which is adopted by all parameter extraction methods examined. Please note that sometimes the literature neglects one or both resistances for simplicity [11,15,17,21,22,55].

The current-voltage equation is given implicitly by

$$I = I_{ph} - I_s \left(e^{\frac{V+IR_s}{a}} - 1 \right) - \frac{V + IR_s}{R_{sh}}. \tag{1}$$

This equation cannot be solved symbolically and necessitates numerical solution, which raises some challenges during the evaluation. As an alternative, an equivalent explicit formulation has appeared lately in the literature, employing the principal branch of the Lambert W function $W_0\{.\}$ [25,56–58]

$$I = \frac{R_{sh}(I_{ph} + I_s) - V}{R_s + R_{sh}} - \frac{a}{R_s} W_0 \left\{ \frac{R_s R_{sh} I_s}{a(R_s + R_{sh})} e^{\frac{R_s R_{sh}(I_{ph}+I_s)+R_{sh}V}{a(R_s+R_{sh})}} \right\} \tag{2}$$

$$V = R_{sh}(I_{ph} + I_s) - (R_s + R_{sh})I - aW_0 \left\{ \frac{R_{sh} I_s}{a} e^{\frac{R_{sh}(I_{ph}+I_s-I)}{a}} \right\}. \tag{3}$$

One can directly find the current for a given value of voltage using (2) or the opposite via (3), which makes the calculation easy and straightforward, in contrast to (1). The Lambert W function is readily available in all computational platforms; more information on the computation of this function are given in Appendix A.

The single-diode model describes any PV generator under uniform operating conditions, from cell to array, after scaling properly the five parameters [43]; it does not apply to non-uniform operation, such as partial shading, mismatched conditions etc., when appropriate extensions to account for the bypass diodes are necessary. Usually, the five parameters are extracted for the PV module, using the module datasheet information as input data. In the general case, however, one can extract the five parameters for any PV generator, from cell to array, when the respective I-V curve measurements are available. To relate the array parameters $[I_{ph,arr}, I_{s,arr}, a_{arr}, R_{s,arr}, R_{sh,arr}]$ to the cell parameters $[I_{ph,cell}, I_{s,cell}, a_{cell}, R_{s,cell}, R_{sh,cell}]$ the following expressions can be used [4,21,34,41,54]:

$$I_{ph,arr} = N_p I_{ph,cell} \tag{4}$$

$$I_{s,arr} = N_p I_{s,cell} \tag{5}$$

$$a_{arr} = N_s a_{cell} \tag{6}$$

$$R_{s,arr} = \frac{N_s}{N_p} R_{s,cell} \tag{7}$$

$$R_{sh,arr} = \frac{N_s}{N_p} R_{sh,cell}, \tag{8}$$

where N_p and N_s are the parallel- and series-connected PV cells within the array. These relations hold true for the other PV structures as well, such as the string or module, properly changing the meaning of the respective terms; e.g., the module-cell relation is given by (4)–(8) if the terms for the array parameters are substituted by the module ones and N_p, N_s meaning is modified to account for the cells within the module, rather than the array. A typical first-quadrant I-V curve (non-negative voltage and current) of a PV module at STC (standard test conditions) is shown in Figure 2, indicating the SC, the OC, and the MPP, as well as the tangent lines at SC and OC. Most of the non-iterative parameter extraction methods require as input data the coordinates of these points $(0, I_{sc})$, $(V_{oc}, 0)$ and (V_{mp}, I_{mp}), as well as the so-called *experimental resistances* R_{sho} and R_{so} that relate to the slope of the tangent lines according to

$$R_{sho} = -\frac{dV}{dI}\bigg|_{SC}, \quad R_{so} = -\frac{dV}{dI}\bigg|_{OC}. \tag{9}$$

The extraction of these slopes from *I-V* curve measurements is discussed in Appendix B. Other data possibly needed by a parameter extraction method is the temperature coefficients of the SC current α_{Isc} and OC voltage β_{Voc}. Please note that these coefficients are used in normalized form (K^{-1}) in this paper (normalized on the nominal SC current and OC voltage, e.g., $\beta_{Voc} = -0.0034$ K^{-1}, $\alpha_{Isc} = +0.0006$ K^{-1}), rather than in absolute form (A/K, V/K), as the normalized coefficients do not considerably differ among PV modules of the same technology which facilitates comparison [59].

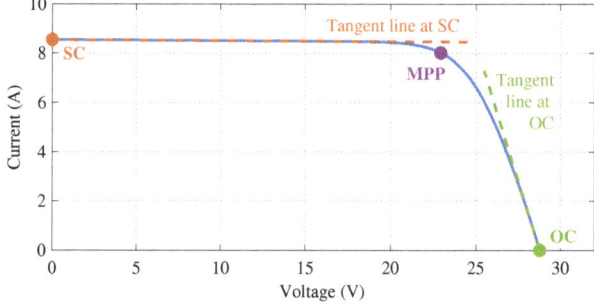

Figure 2. Typical *I-V* characteristic curve of a PV module in the first quadrant.

3. The Non-Iterative Parameter Extraction Methods

A rigorous review reveals that there are *seventeen* different non-iterative methods available for the parameter extraction of the *single-diode PV model*, given in chronological order and denoted by the name of their respective main author in Table 1. These are the most clearly described methods in the literature that are applicable to any operating conditions; the ones intended *only* for STC are not included in this comparative assessment. It is worth noting that some of these methods are designed for single/multi-crystalline silicon technologies, assuming the silicon energy gap and a diode factor close to 1, but in this study, they are applied to all six PV technologies to assess their universal performance maintaining consistency with the original studies. Table 1 also shows the number of evaluation steps, the required input data and whether the datasheet information suffices, or additional measurements are needed. In the rest of this section, the evaluation steps of these methods are briefly described using the common nomenclature of Section 2; it is shown that there is significant overlap among the alternatives, sharing several equations in the exact or very similar form.

3.1. Phang [1984]

Historically, the first attempts to extract the PV model parameters appeared in the 1960s: they employed semilogarithmic plots of the *I-V* characteristic to extract some of the parameters and required experienced users. These techniques were evolved later to complete and easy-to-use parameter extraction methods based on explicit equations, rather than plotted curves; the first such non-iterative method was published by Phang et al. in 1984 [14], still performing quite decently as discussed later. The same authors published later a comparative study [60] and a seven-parameters extraction method [61], but [14] remains the original publication adopted in several studies in the literature, such as in [50,62]. First, R_{sho} and R_{so} are extracted from the *I-V* curve slopes at SC and OC (see Appendix B), and then the five parameters are calculated by evaluating the following equations in this order:

$$R_{sh} = R_{sho} \tag{10}$$

$$a = \frac{V_{mp} - V_{oc} + R_{so}I_{mp}}{\ln\left[\frac{I_{sc} - I_{mp} - V_{mp}/R_{sh}}{I_{sc} - V_{oc}/R_{sh}}\right] + \frac{I_{mp}}{I_{sc} - V_{oc}/R_{sh}}} \tag{11}$$

$$I_s = \left(I_{sc} - \frac{V_{oc}}{R_{sh}}\right) e^{-\frac{V_{oc}}{a}} \tag{12}$$

$$R_s = R_{so} - \frac{a}{I_s} e^{-\frac{V_{oc}}{a}} \tag{13}$$

$$I_{ph} = I_{sc}\left(1 + \frac{R_s}{R_{sh}}\right) + I_s\left(e^{\frac{I_{sc}R_s}{a}} - 1\right). \tag{14}$$

Table 1. Main attributes of the 17 non-iterative parameter extraction methods.

No	Ref.	Method	Year	Number of Parameters	Eval Steps	Datasheet Sufficient?	Input Data
1	[14]	Phang	1984	5	7		$I_{sc}, V_{oc}, I_{mp}, V_{mp}, R_{sho}, R_{so}$
2	[15]	Sera	2008	4	4	✓	$I_{sc}, V_{oc}, I_{mp}, V_{mp}$
3	[16]	Saleem	2009	5	7		$I_{sc}, V_{oc}, V_{mp}, I_{60}, V_{60}$
4	[17]	Saloux	2011	3	3	✓	$I_{sc}, V_{oc}, I_{mp}, V_{mp}$
5	[12]	Accarino	2013	5	6	✓	$I_{sc}, V_{oc}, I_{mp}, V_{mp}, \alpha_{Isc}, \beta_{Voc}, V_{oc0}$
6	[18]	Khan	2013	5	7		$I_{sc}, V_{oc}, I_{mp}, V_{mp}, R_{sho}, R_{so}$
7	[19]	Cubas1	2014	5	8		$I_{sc}, V_{oc}, I_{mp}, V_{mp}, R_{sho}$
8	[19]	Cubas2	2014	5	8	✓	$I_{sc}, V_{oc}, I_{mp}, V_{mp}$
9	[56]	Cubas3	2014	5	9	✓	$I_{sc}, V_{oc}, I_{mp}, V_{mp}$
10	[20]	Bai	2014	5	11	✓	$I_{sc}, V_{oc}, I_{mp}, V_{mp}$
11	[21]	Aldwane	2014	4	4	✓	$I_{sc}, V_{oc}, I_{mp}, V_{mp}$
12	[22]	Cannizzaro	2014	4	8 or 11	✓	$I_{sc}, V_{oc}, I_{mp}, V_{mp}$
13	[23]	Toledo	2014	5	13		$I_{sc}, V_{oc}, I_{mp}, V_{mp}, I_{xx}, R_{sho}$
14	[24]	Louzazni	2015	5	7		$I_{sc}, V_{oc}, I_{mp}, V_{mp}, R_{sho}, R_{so}$
15	[25]	Batzelis	2016	5	8	✓	$I_{sc}, V_{oc}, I_{mp}, V_{mp}, \alpha_{Isc}, \beta_{Voc}, V_{oc0}$
16	[26]	Hejri	2016	5	5	✓	$I_{sc}, V_{oc}, I_{mp}, V_{mp}$
17	[27]	Senturk	2017	5	7	✓	$I_{sc}, V_{oc}, I_{mp}, V_{mp}$

3.2. Sera [2008]

The method of Sera et al. [15], also adopted by Khezzar et al. [55], is a four-parameter model that neglects R_{sh}:

$$I_{ph} = I_{sc} \tag{15}$$

$$a = \frac{2V_{mp} - V_{oc}}{\ln\left(\frac{I_{sc} - I_{mp}}{I_{sc}}\right) + \frac{I_{mp}}{I_{sc} - I_{mp}}} \tag{16}$$

$$R_s = \frac{a \ln\left(\frac{I_{sc} - I_{mp}}{I_{sc}}\right) + V_{oc} - V_{mp}}{I_{mp}} \tag{17}$$

$$I_s = I_{sc} e^{-\frac{V_{oc}}{a}}. \tag{18}$$

It is worth noting that a typo found in [15] has been corrected in (16), while (18) is somewhat similar to the Phang's (12).

3.3. Saleem [2009]

The method of Saleem [16] is based on an alternative formulation of the PV equation as a power law function. It requires two additional points as input data: at voltage equal to 60% of V_{oc} ($0.6V_{oc}, I_{60}$) and at current equal to 60% of I_{sc} ($V_{60}, 0.6I_{sc}$). First, the auxiliary parameters γ and m are calculated:

$$\gamma = \frac{\frac{I_{60}}{I_{sc}} - 0.4}{0.6}, \quad m = \frac{\log\left(\frac{0.4-(1-\gamma)\frac{V_{60}}{V_{oc}}}{\gamma}\right)}{\log \frac{V_{60}}{V_{oc}}}, \quad (19)$$

and then the five parameters are found by

$$a = \frac{V_{oc}}{m} \frac{0.77m(1 - \frac{V_{mp}}{V_{oc}}) - 1}{0.77m \ln \frac{V_{oc}}{V_{mp}} - 1} \quad (20)$$

$$R_s = \frac{V_{oc}}{0.6\gamma m I_{sc}} \left(1 - \frac{am}{V_{oc}}\right) - 0.1 \quad (21)$$

$$I_s = \gamma I_{sc} e^{-\frac{V_{oc}}{a}} \quad (22)$$

$$R_{sh} = \frac{V_{oc}}{I_{sc}} \frac{1}{1 - \gamma - \frac{\gamma}{0.6} \exp\left[\frac{(0.4+0.6\gamma)I_{sc}R_s - 0.4V_{oc}}{a}\right]} \quad (23)$$

$$I_{ph} = I_{sc} \left(1 + \frac{R_s}{R_{sh}}\right). \quad (24)$$

Note the logarithm with a base of 10 in (19) and the natural logarithm in (20).

3.4. Saloux [2011]

An explicit PV model based on the ideal single-diode equivalent (no resistances) is presented in [17]. Saloux provides some equations to determine the three parameters at STC, but it seems that they are applicable to other conditions as well, as slightly manipulated here:

$$a = \frac{V_{mp} - V_{oc}}{\ln\left(1 - \frac{I_{mp}}{I_{sc}}\right)} \quad (25)$$

$$I_{ph} = I_{sc} \quad (26)$$

$$I_s = \frac{I_{sc}}{e^{\frac{V_{oc}}{a}} - 1}. \quad (27)$$

Apparently, the expressions for I_{ph} and I_s are almost the same as (15) and (18) from Sera's model.

3.5. Accarino [2013]

This is the first extraction method to employ the Lambert W function [12]. First a is calculated via

$$a = \frac{\left(\beta V_{oc} - \frac{1}{T_0}\right) V_{oc0}}{\alpha I_{sc} - \frac{3}{T_0} - \frac{E_g}{kT_0^2}} \frac{T}{T_0}, \quad (28)$$

where V_{oc0} is the nominal OC voltage (STC), $T_0 = 298.15$ K is the nominal temperature, $E_g = 1.8e - 19$ J the energy gap of silicon and $k = 1.38e - 23$ J/K^2 the Boltzmann constant. Also note that the temperature coefficients are in normalized form. Then, I_{ph} and I_s are found by (15) and (18) from Sera's model. Finally, the auxiliary parameter x is evaluated (after correcting a typo in [12]), used subsequently to calculate R_s and R_{sh}:

$$x = W_0 \left\{ \frac{V_{mp}(2I_{mp} - I_{ph})}{aI_s} e^{\frac{V_{mp}(V_{mp}-2a)}{a^2}} \right\} + 2\frac{V_{mp}}{a} - \left(\frac{V_{mp}}{a}\right)^2 \tag{29}$$

$$R_s = \frac{xa - V_{mp}}{I_{mp}}, \quad R_{sh} = \frac{xa}{I_{ph} - I_{mp} - I_s(e^x - 1)}. \tag{30}$$

For the computation of the principal branch of the Lambert W function $W_0\{.\}$, please see Appendix A. It is worth noting that this method was applied in [12] to single/multi-crystalline silicon modules assuming the energy gap of silicon, thus it is also applied here using the same E_g value for all technologies to be consistent with original study.

3.6. Khan [2013]

The model of Khan requires R_{sho} and R_{so} as additional inputs [18]. First, R_s, a and I_s are calculated

$$R_s = R_{so} - \frac{V_{mp} - V_{oc} + R_{so}I_{mp}}{I_{mp} + I_{sc}\ln\left(1 - \frac{I_{mp}}{I_{sc}}\right)} \tag{31}$$

$$a = \frac{V_{mp} - V_{oc} + R_s I_{mp}}{\ln\left(1 - \frac{I_{mp}}{I_{sc}}\right)} \tag{32}$$

$$I_s = \frac{a}{R_{so} - R_s} e^{-\frac{V_{oc}}{a}}, \tag{33}$$

and then R_{sh} and I_{ph} are found via Phang's (10) and (14).

3.7. Cubas1 [2014]

Cubas et al. propose one numerical and two non-iterative parameter extraction methods in [19]. The first non-iterative approach requires measurement of R_{sho} to calculate R_s through the auxiliary parameters A and B

$$A = [V_{mp} + (I_{mp} - I_{sc})R_{sho}]\ln\left(\frac{V_{mp} + (I_{mp} - I_{sc})R_{sho}}{V_{oc} - I_{sc}R_{sho}}\right), \quad B = V_{mp} - R_{sho}I_{mp} \tag{34}$$

$$R_s = \frac{A - B\,V_{mp}}{A + B\,I_{mp}} + \frac{B}{A + B} \frac{V_{oc}}{I_{mp}}, \tag{35}$$

and evaluate a, R_{sh} and I_s through

$$a = \frac{(V_{mp} - I_{mp}R_s)[V_{mp} + (I_{mp} - I_{sc})R_{sho}]}{V_{mp} - I_{mp}R_{sho}} \tag{36}$$

$$R_{sh} = R_{sho} - R_s \tag{37}$$

$$I_s = \left[I_{sc}\left(1 + \frac{R_s}{R_{sh}}\right) - \frac{V_{oc}}{R_{sh}}\right] e^{-\frac{V_{oc}}{a}}. \tag{38}$$

Finally, I_{ph} is found through Saleem's (24).

3.8. Cubas2 [2014]

The second non-iterative alternative proposed in [19] employs an empirical estimation of R_{sho} derived from [63], in place of the SC slope measurement:

$$R_{sho} = 34.49692 \frac{V_{oc}}{I_{sc}}. \tag{39}$$

This renders the method reliant only on datasheet information. The remaining parameters are calculated via the equations of the previous alternative (24), (34)–(36) and (38), except that R_{sh} is now found through:

$$R_{sh} = \frac{(V_{mp} - I_{mp}R_s)[V_{mp} - R_s(I_{sc} - I_{mp}) - a]}{(V_{mp} - I_{mp}R_s)(I_{sc} - I_{mp}) - aI_{mp}} \tag{40}$$

3.9. Cubas3 [2014]

The same authors published shortly after a similar study based on the Lambert W function [56], which requires the diode factor n as an input, set to $n = 1.1$ for the silicon cells under study:

$$a = \frac{nN_s kT}{q}. \tag{41}$$

Four auxiliary parameters A, B, C and D are used

$$A = \frac{a}{I_{mp}}, \qquad B = -\frac{V_{mp}(2I_{mp} - I_{sc})}{[V_{mp}I_{sc} + V_{oc}(I_{mp} - I_{sc})]}, \tag{42}$$

$$C = -\frac{2V_{mp} - V_{oc}}{a} + \frac{V_{mp}I_{sc} - V_{oc}I_{mp}}{[V_{mp}I_{sc} + V_{oc}(I_{mp} - I_{sc})]}, \qquad D = \frac{V_{mp} - V_{oc}}{a} \tag{43}$$

to calculate R_s by

$$R_s = A\left[W_{-1}\left\{Be^C\right\} - (D+C)\right]. \tag{44}$$

Please note that $W_{-1}\{.\}$ in (44) is the lower branch of the Lambert W function, rather than the more commonly used principal branch $W_0\{.\}$; the calculation formula is given in Appendix A. The remaining parameters R_{sh}, I_s and I_{ph} are found by the previous alternatives' Equations (40), (38) and (24). Please note that in this paper, the same diode factor is used for all PV technologies.

3.10. Bai [2014]

The Bai method [20] requires the slopes at SC and OC as inputs, which however are estimated rather than measured. First, a simplified four-parameter model is employed to derive $[I_{ph4}, a_4, R_{s4}, I_{s4}]$ through Sera's (15)–(18). Then, the SC and OC slopes, or equivalently R_{sho} and R_{so} are calculated by

$$R_{sho} = \frac{a_4 \ln\left(\frac{0.5(I_{sc} - I_{mp})}{I_{s4}} + 1\right) - 0.5(I_{sc} + I_{mp})R_{s4}}{0.5(I_{sc} - I_{mp})} \tag{45}$$

$$R_{so} = -\frac{a_4 \ln\left(\frac{I_{sc} - 0.5 I_{mp}}{I_{s4}} + 1\right) - 0.5 I_{mp} R_{s4} - V_{oc}}{0.5 I_{mp}}. \tag{46}$$

Thereafter, R_s is calculated by (47) below, R_{sh} and I_{ph} through Cubas1's (37) and Saleem's (24) respectively, and finally a and I_s by (48) and (49) below.

$$R_s = \frac{V_{mp}(-R_{so} + R_{sho})[-R_{sho}(I_{sc} - I_{mp}) + V_{mp}] + R_{so}(-R_{sho}I_{mp} + V_{mp})(-R_{sho}I_{sc} + V_{oc})}{I_{mp}(-R_{so} + R_{sho})[-R_{sho}(I_{sc} - I_{mp}) + V_{mp}] + (-R_{sho}I_{mp} + V_{mp})(-R_{sho}I_{sc} + V_{oc})} \quad (47)$$

$$a = (-R_{so} + R_s)\frac{-R_{sho}I_{sc} + V_{oc}}{-R_{so} + R_{sho}} \quad (48)$$

$$I_s = \frac{I_{ph} - \frac{V_{oc}}{R_{sh}}}{e^{\frac{V_{oc}}{a}} - 1}. \quad (49)$$

Evidently, (49) is almost identical to Phang's (12).

3.11. Aldwane [2014]

This is a four-parameter model (R_{sh} is neglected) [21], which is very similar to Sera's method except for a small difference in a:

$$a = \frac{2V_{mp} - V_{oc}}{\ln\left(\frac{I_{sc} - I_{mp}}{I_{sc}}\right) + \frac{I_{sc}}{I_{sc} - I_{mp}}}. \quad (50)$$

The remaining parameters I_{ph}, R_s and I_s are found via the exact same Sera's Equations (15), (17) and (18) respectively.

3.12. Cannizzaro [2014]

Cannizzaro et al. developed their method initially in [11], thereafter completed in [22]. This approach is based on the assumption that the five-parameter model can always be simplified to a four-parameter model, neglecting either R_s or R_{sh} according to the series to parallel ratio (SPR):

$$SPR = \frac{1 - \gamma_i}{e^{-r}}, \quad \text{where:} \quad \gamma_i = \frac{I_{mp}}{I_{sc}}, \quad \gamma_v = \frac{V_{mp}}{V_{oc}}, \quad r = \frac{\gamma_i(1 - \gamma_v)}{\gamma_v(1 - \gamma_i)}. \quad (51)$$

If $SPR \geq 1$, the shunt resistance is neglected, and the series resistance is given by

$$R_s = \frac{V_{oc}}{I_{sc}}\frac{\gamma_v(1 - \gamma_i)\ln(1 - \gamma_i) + (1 - \gamma_v)}{\gamma_i(1 - \gamma_i)\ln(1 - \gamma_i) + \gamma_i}, \quad R_{sh} = \infty. \quad (52)$$

Otherwise, if $SPR < 1$:

$$R_s = 0, \quad R_{sh} = \frac{V_{oc}}{I_{sc}}\frac{\lambda_2 w + \lambda_1}{w + \lambda_1}, \quad (53)$$

$$\text{where } \lambda_1 = \frac{1 - \gamma_v}{1 - \gamma_i}\frac{2\gamma_i - 1}{\gamma_i + \gamma_v - 1}, \quad \lambda_2 = \frac{\gamma_v}{1 - \gamma_i}, \quad w = W_{-1}\left\{-SPR\lambda_1 e^{-\lambda_1}\right\}. \quad (54)$$

The term $W_{-1}\{.\}$ in (54) is the lower branch of the Lambert W function, approximated in this paper by the formula used by Cannizzaro et al. in [22] (more details in Appendix A). Having the two resistances permits evaluation of a through

$$a = \frac{V_{mp} - V_{oc} + I_{mp}R_s}{\ln\left[\frac{(I_{sc} - I_{mp})(1 + \frac{R_s}{R_{sh}}) - \frac{V_{mp}}{R_{sh}}}{I_{sc}(1 + \frac{R_s}{R_{sh}}) - \frac{V_{oc}}{R_{sh}}}\right]}. \quad (55)$$

Evidently, (55) is reduced to Khan's (32) when $SPR \geq 1$ ($R_{sh} = \infty$). Finally, I_{ph} is found via Saleem's (24) and I_s through Phang's (12).

3.13. Toledo [2014]

Apart from R_{sho}, the method in [23] uses as inputs the coordinates of the SC and another three operating points $(V_1, I_1), (V_2, I_2), (V_3, I_3)$ evenly distributed at the right-hand side of the I-V curve. There is some flexibility on how to select these points; here the MPP, OC and XX ($V_{xx} = \frac{V_{mp}+V_{oc}}{2}$—notation introduced by SANDIA laboratories [64]) are used. The method employs five auxiliary parameters $[A, B, C, D, E]$ found via the following steps. First, the sum $A + B$ and E are derived to calculate the function f_i for the three operating points (V_i, I_i):

$$A + B = I_{sc}, \quad E = 1/R_{sho}, \quad f_i = \ln(A + B - EV_i - I_i) \text{ for } i = 1,2,3. \tag{56}$$

Using this information, the rest of the parameters are found by

$$D = \exp\left[\frac{(f_1 - f_2)(V_2 - V_3) - (f_2 - f_3)(V_1 - V_2)}{(I_1 - I_2)(V_2 - V_3) - (I_2 - I_3)(V_1 - V_2)}\right], \quad C = \exp\left[\frac{f_2 - f_3 - (I_2 - I_3)\ln D}{V_2 - V_3}\right] \tag{57}$$

$$B = \exp(f_1 - V_1 \ln C - I_1 \ln D), \quad A = I_{sc} - B. \tag{58}$$

Finally, the five parameters are retrieved:

$$I_{ph} = \frac{A \ln C}{\ln C - E \ln D} \tag{59}$$

$$I_s = \frac{B \ln C}{\ln C - E \ln D} \tag{60}$$

$$a = \frac{1}{\ln C} \tag{61}$$

$$R_s = \frac{\ln D}{\ln C} \tag{62}$$

$$R_{sh} = \frac{1}{E} - \frac{\ln D}{\ln C}. \tag{63}$$

3.14. Louzazni [2015]

The main objective of Louzazni et al. in [24] is to present an alternative formulation of the PV model equation, in which the five parameters are found as follows. First, the experimental resistances are approximated by the respective slopes: R_{sh} via Phang's (10) and R_s via

$$R_s = R_{so}. \tag{64}$$

Thereafter, a is calculated via Phang's (11), I_{ph} via Saleem's (24) and I_s via Sera's (18).

3.15. Batzelis [2016]

The Batzelis method [25] introduces a coefficient δ_0 at STC found through the temperature coefficients $\alpha_{Isc}, \beta_{Voc}$, used afterwards to derive the auxiliary parameters δ and w:

$$\delta_0 = \frac{1 - \beta_{Voc} T_0}{50.1 - \alpha_{Isc} T_0} \tag{65}$$

$$\delta = \delta_0 \frac{V_{oc0}}{V_{oc}} \frac{T}{T_0} \tag{66}$$

$$w = W_0\left\{e^{\frac{1}{\delta}+1}\right\}. \tag{67}$$

Please note that $\alpha_{Isc}, \beta_{Voc}$ are normalized, T, T_0 are in Kelvin degrees and the $W_0\{.\}$ is the principal branch of the Lambert W function (please see Appendix A). The term 50.1 in the denominator of δ_0 incorporates the Boltzmann's constant and the energy gap of silicon; this coefficient is used as it is in this paper for all PV technologies examined, in order to be consistent with the original paper. Using these coefficients, the five parameters are found by

$$a = \delta V_{oc} \tag{68}$$

$$R_s = \frac{a(w-1) - V_{mp}}{I_{mp}} \tag{69}$$

$$R_{sh} = \frac{a(w-1)}{I_{sc}(1 - \frac{1}{w}) - I_{mp}} \tag{70}$$

$$I_{ph} = \left(1 + \frac{R_s}{R_{sh}}\right) I_{sc} \tag{71}$$

$$I_s = I_{ph} e^{-\frac{1}{\delta}}. \tag{72}$$

3.16. Hejri [2016]

The main purpose of the Hejri method [26] is to approximate the five parameters to be used as initial values in a numerical solution algorithm. First, a is found via Sera's (16), then R_s and R_{sh} by

$$R_s = \frac{V_{mp}}{I_{mp}} - \frac{\frac{2V_{mp} - V_{oc}}{I_{sc} - I_{mp}}}{\ln\left(\frac{I_{sc} - I_{mp}}{I_{sc}}\right) + \frac{I_{mp}}{I_{sc} - I_{mp}}} \tag{73}$$

$$R_{sh} = \sqrt{\frac{R_s}{\frac{I_{sc}}{a} \exp\left(\frac{R_s I_{sc} - V_{oc}}{a}\right)}}. \tag{74}$$

Finally, I_{ph} and I_s are calculated through Sera's (15) and (18) respectively. It is worth noting that during the derivations of these expressions, it has been assumed that $R_{sho} = R_{sh}$ to avoid measuring the SC slope.

3.17. Senturk [2017]

This method is intended for STC [27], but it seems to be applicable to other conditions as well. First, an arbitrary value of $n = 1.2$ is assumed for the diode factor, thus calculating a by (41) as in Cubas3. Then, the experimental resistances are approximated by

$$R_{sho} = \frac{V_{mp}}{I_{sc} - I_{mp}} \tag{75}$$

$$R_{so} = \frac{V_{oc} - V_{mp}}{2I_{mp}}. \tag{76}$$

Thereafter, I_{ph} and I_s are calculated by

$$I_{ph} = \frac{R_{so} + R_{sho}}{R_{sho}} I_{sc} \tag{77}$$

$$I_s = \frac{I_{ph} - \frac{V_{oc}}{R_{sho}}}{e^{\frac{V_{oc}}{a}} - 1}, \tag{78}$$

and finally R_s via Phang's (13) and R_{sh} by

$$R_{sh} = \frac{V_{mp} + I_{mp}R_s}{I_{ph} - I_{mp} - I_s\left(e^{\frac{V_{mp}+I_{mp}R_s}{a}} - 1\right)}. \tag{79}$$

This method too is designed for single/multi-crystalline silicon modules but is applied here as it is to all technologies under consideration.

4. Comparative Assessment

All 17 methods of Section 3 are implemented and assessed in MATLAB 2017b, in a PC with a 6-core 3.5-GHz CPU at sequential computational mode (not parallel). Each method is denoted by the name of the respective main author, as shown in Table 1. The dataset used for the evaluation was kindly provided by the NREL and contains 1,025,665 I-V curves and other measurements from 22 PV modules of different technology; each curve is recorded in about 180–200 samples. These measurements are classified into 6 main PV technologies here, as shown in Table 2.

The irradiance and temperature distribution of this dataset is shown in Figure 3a,b, the former varying within [20, 1440] W/m² and the latter within [−19, +73] °C. A large number of I-V curves were recorded under low and very low-irradiance conditions, which are known to be challenging for the single-diode model and therefore for the respective parameter extraction methods. More information on this dataset may be found in [42]. The comparative assessment that follows is first performed for all 17 methods, repeated afterwards for those that rely only on datasheet information and for those that require the SC or/and OC slope measurements.

Table 2. Dataset of I-V curves used in the comparative assessment provided by NREL [42].

PV Module	Technology	No of Curves
xSi12922	c-Si	94,109
xSi11246		
mSi460BB	mc-Si	280,042
mSi460A8		
mSi0251		
mSi0247		
mSi0188		
mSi0166		
HIT05667	HIT	93,530
HIT05662		
CIGS8-001	CIGS	182,994
CIGS39017		
CIGS39013		
CIGS1-001		
CdTe75669	CdTe	93,287
CdTe75638		
aSiTriple28325	a-Si	281,703
aSiTriple28324		
aSiTandem90-31		
aSiTandem72-46		
aSiMicro03038		
aSiMicro03036		
TOTAL	6	1,025,665

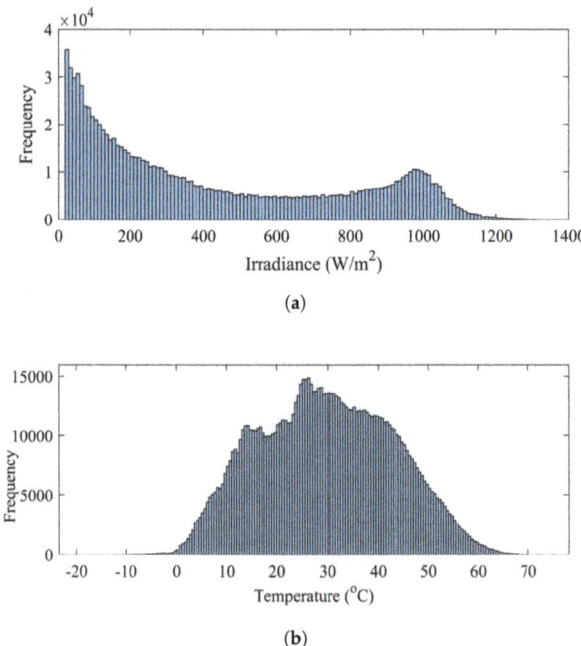

Figure 3. Histograms of (**a**) irradiance and (**b**) temperature distribution of the study-case dataset.

4.1. All Seventeen Methods

The full picture in the performance of the 17 methods is given in Table 3 cumulatively for all PV technologies. The methods whose name start with an "*" are the ones that rely only on datasheet information (see Table 1), while bold highlighting indicates the lowest value in the respective columns. The assessment is carried out using three main criteria: accuracy, robustness, and complexity. The metric for the accuracy is the current Root Mean Square Error (RMSE) of the reconstructed and measured I-V curve, which indicates how well the characteristic produced using the extracted parameters matches the sampled one; Table 1 gives the mean/max values of the absolute (A) and normalized (on nominal I_{sc}—%) RMSE, as both metrics are used in the literature. The robustness is assessed on the number of scenarios out of the 1 million dataset that led to irregular parameters (negative or complex parameters) or to execution failure, and the computational performance is measured on the basis of total execution time and core execution time (excluding the I-V curve calculations, such as extraction of SC slope).

It is evident that the various methods perform very differently in every aspect of this comparison. The best mean (absolute and normalized) RMSE is shown by the Toledo method, which however exhibits moderate worst-case performance. Conversely, Cannizzaro and Cubas3 (only for absolute RMSE) provide the lowest max RMSE, but present much higher average errors than Toledo. The majority of the methods have thousands of irregularities in the extracted parameters, mainly negative but also complex values, which may be undesirable for some applications. The fewer irregularities are shown by the very simple Saloux method that is based on the three-parameter model, which yields also the lowest computational cost; still, the accuracy of this method seems to be below par. It is interesting that so many methods suffer from hundreds of thousands of execution failures, failing to reproduce a meaningful I-V curve with the extracted parameters; only the Saloux, Batzelis, and Senturk approaches guarantee 100% successful execution. These results are discussed in more detail in the remaining of this section.

Table 3. Performance of all 17 parameters extraction methods.

Method	Accuracy (RMSE)				Robustness		Complexity	
	Absolute (A)		Normalized (%)				Execution Time (s)	
	Mean	Max	Mean	Max	Irregularities	Failures	Total	Core
Phang	0.016	4.210	0.38	69.4	387,440	9952	13.1	1.3
* Sera	0.026	2.471	0.74	49.6	794,670	52	1.3	1.3
Saleem	0.018	2.255	0.47	45.3	22,810	24	10.7	2.1
* Saloux	0.029	1.271	0.79	25.5	2	0	**1.2**	1.2
* Accarino	0.034	0.880	1.09	17.7	2804	25	1.5	1.5
Khan	0.043	4.201	1.19	69.3	511,210	8242	13.2	1.4
Cubas1	0.011	2.445	0.36	49.1	529,530	2445	8.2	**1.1**
* Cubas2	0.026	2.433	0.64	48.8	626,150	53	1.3	1.3
* Cubas3	0.024	**0.430**	0.83	13.0	1878	224	2.0	2.0
* Bai	0.032	2.544	0.93	51.1	6390	37	1.7	1.7
* Aldwane	0.021	1.032	0.59	20.7	416,650	29	1.3	1.3
* Cannizzaro	0.013	**0.431**	0.39	**11.3**	100	28	2.1	2.1
Toledo	**0.007**	1.119	**0.20**	97.7	451,220	392	14.0	2.5
Louzazni	0.122	1.449	3.43	27.0	64,194	30,612	12.5	**1.1**
* Batzelis	0.028	0.932	0.87	18.7	412	0	1.5	1.5
* Hejri	0.231	6.761	7.21	129.4	794,670	14,057	5.4	5.4
* Senturk	0.034	1.285	1.09	25.8	27,847	0	1.2	1.2

The asterisk "*" denotes methods that rely only on datasheet information. Bold font indicates the lowest value in the respective column.

4.1.1. Accuracy

The mean RMSE varies in the range of ~10–50 mA (0.2–1.2%) for all methods, except for Louzazni and Hejri that perform much worse; the Toledo method yields the best average accuracy with a mean error of 7 mA (0.2%). To get an idea of how good this performance is, it is worth noting that the lowest mean absolute RMSE reported for numerical/iterative methods for the same NREL dataset is in the range of 2–3 mA [34]: most of the non-iterative methods exhibit a mean error of about one order of magnitude larger.

For the worst-case accuracy, however, the errors recorded in Table 3 are much higher; only the Accarino, Cubas3, Cannizzaro, and Batzelis methods present a max RMSE of less than 1 A (20%) for all one million scenarios. It is interesting that Toledo yields much higher max errors compared to Cannizaro, which indicates that best average accuracy does not entail best worst-case accuracy as well.

The performance for the various PV technologies is graphically illustrated in Figure 4. Each stacked bar in Figure 4a corresponds to the sum of the mean normalized RMSE for all six technologies. It is evident that the largest errors arise for the CIGS (purple) and a-Si (light blue) technologies in all methods, which indicates that these are the most challenging type to describe through the single-diode model; the other technologies are pretty much the same from the modeling perspective.

Figure 4b shows the PV technology that corresponds to the max normalized RMSE for each method: the largest worst-case errors are found in CIGS (purple), CdTe (green) and a-Si (light blue) technologies, but also several methods yield their max RMSE surprisingly at the mono-crystalline c-Si technology (dark blue). This last observation indicates that the Achilles' heel of these methods is primarily the simplifications in their core, and secondarily the appropriateness of the single-diode model for different PV technologies.

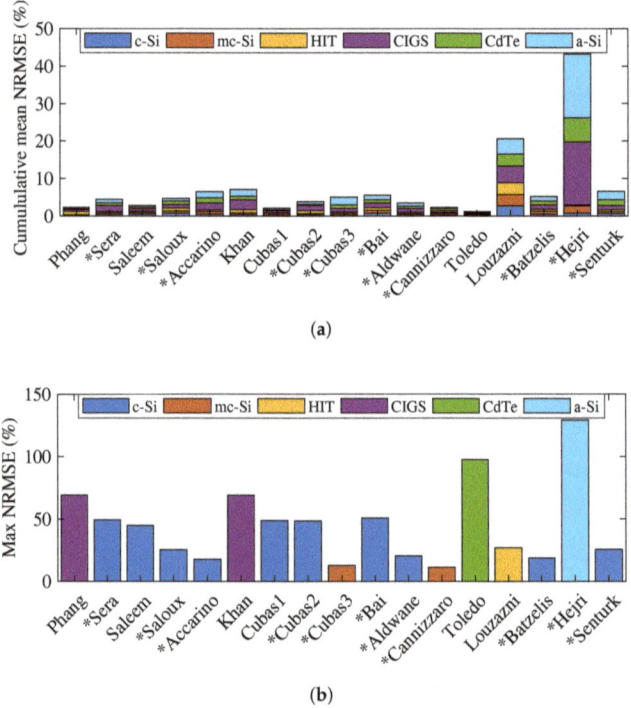

Figure 4. Accuracy of the 17 methods. (**a**) Cumulative mean normalized RMSE and (**b**) max normalized RMSE for the various PV technologies.

An indication of the most accurate method for each of the six PV technologies is given in Table 4. Evidently, Toledo proves the best option in all six technologies when focusing on average accuracy, as discussed before. For the worst-case accuracy, the picture is more complicated having a different method for each type of modules, having Cannizzaro prevalent and Toledo entirely absent in this column; these results indicate that there is no "best" method to be blindly adopted. This is expected, as all methods are based on the single-diode PV model which has been developed for the single/multi-crystalline silicon technologies, thus setting a limit to the accuracy that can be achieved in other types of modules. What determines the most favorably performing method in each case is how valid the adopted assumptions are in the different technologies. For example, it seems that neglecting either the series or the shunt resistance leads to good worst-case accuracy for the Cannizzaro method in the silicon technologies (c-, mc- and a-Si), whereas a five-parameter model is needed for the other technologies.

Table 4. Most accurate methods for each PV technology.

	Best Average Accuracy		Best Worst-Case Accuracy	
PV Technology	Method	Mean NRMSE (%)	Method	Max NRMSE (%)
c-Si	Toledo	0.10	* Cannizzaro	5.6
mc-Si	Toledo	0.20	* Cannizzaro	11.3
HIT	Toledo	0.23	Saleem	2.9
CIGS	Toledo	0.25	Cubas1	2.5
CdTe	Toledo	0.17	Phang	2.4
a-Si	Toledo	0.24	* Cannizzaro	5.4

The asterisk "*" denotes methods that rely only on datasheet information.

4.1.2. Robustness

The aspect of robustness has not been studied in a quantitative manner for the non-iterative methods in the literature before. However, during this investigation it was found that it is common to get negative or complex values for the five parameters from all methods, which may be inappropriate for some applications, or even failing to reconstruct a meaningful *I-V* curve sometimes. In this paper, the robustness of the case study methods is assessed through the number of irregularities found in the parameters and the failures to produce an acceptable *I-V* curve with the extracted parameters.

There is no commonly accepted norm in the literature on whether the five parameters should be restricted to real positive numbers or should be allowed to get negative or complex values. Some studies support the former approach to get parameters with physical meaning [4,26], whereas others do not adhere to this restriction if this leads to better results [15,30]. An example of the latter approach is model-based Maximum Power Point Tracking (MPPT) algorithms, where the five parameters must be extracted easily and non-physical values do not matter if the estimated *P-V* curve is a good approximation of the actual one. To investigate this aspect, a set of parameters that contain at least one negative or complex value is marked as "irregular" here; thus, the *irregularities* shown in Table 3 correspond to the number of scenarios that led to irregular parameters. It is worth noting that all methods returned finite values for all 1 million cases (there was no Inf or NaN in any of the parameters). Table 3 shows that most methods present thousands of parameter irregularities, some of those at almost half of the total scenarios. The fewer irregularities are found in Saloux, Cannizzaro, and Batzelis (less than 0.1%), the former yielding irregular results in only 2 cases; this is probably because of the three-parameter model adopted therein, which neglects the two resistances that most frequently get irregular values.

The distribution of irregularities among the six PV technologies is depicted in Figure 5a: there is no "susceptible" technology, as the number of anomalies is approximately proportional to the number of each technology's scenarios (see Table 2). The type of irregularities are shown in Figure 5b: the majority of the anomalies are just negative values (most often in the resistances), whereas a very small number of complex values have been also observed in most methods; Hejri is an exception yielding a large number of both negative (series resistance) and complex (shunt resistance) parameters, which is due to the assumptions and formulation adopted. For all methods, the complex values come from a negative argument in square root or logarithmic functions.

Still, the irregular parameters are not necessarily an undesirable feature; for some applications, the nature of parameter values do not matter, if they lead to a good match between the produced *I-V* curve and the measured characteristic. To quantify this appropriateness and meaningfulness, we consider here as an *execution failure* when the *I-V* curve contains infinite or NaN (not-a-number) values, or the absolute RMSE exceeds the respective I_{sc} (max current value). Table 3 reveals that some of the methods fail thousands of times, mainly in the HIT and CIGS technologies as shown in Figure 5c; for some methods, this is related to the SC slope extraction, especially in distorted *I-V* curves as further discussed in Section 4.3. This drawback seriously undermines the credibility of these techniques. Only the Saloux, Batzelis and Senturk alternatives fail exactly zero times, which renders them 100% credible for any PV technology and operating conditions.

4.1.3. Computational Complexity

The low complexity and execution time is the main reason to resort to a non-iterative parameter extraction method. The number of evaluation steps shown in Table 1 is only indicative of the computational performance, as each step may widely differ in terms of execution time. Reasons for higher times may be the slopes extraction or calculations with complex numbers. In general, the methods that need to perform calculations on the *I-V* curve measurements (e.g., extraction of the SC slope or locating a specific operating point) carry an additional computational burden. To quantify this cost, the execution time in Table 3 is recorded twice: as a total and excluding this overhead (denoted as *core* cost).

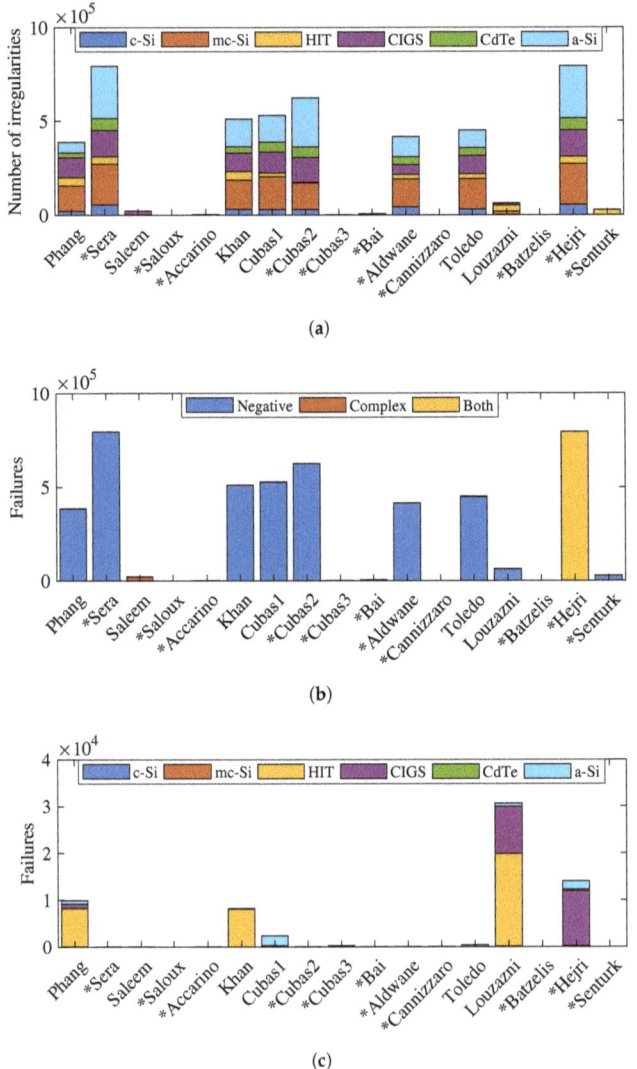

Figure 5. Robustness of the 17 methods. (**a**) Number and (**b**) types of irregularities. (**c**) Execution failures.

Table 3 shows that for all 1 million curves, the total execution time is between 8–14 s for the methods that involve curve calculations (the ones without an "*") and 1–2 s for those that rely only on datasheet information, except for Hejri which exhibits a 5.4 s total time due to the large number of calculations with complex numbers. A graphical illustration of this performance is given in Figure 6a: the increased cost of Phang, Khan, Cubas1, Toledo and Louzazni is due to the extraction of the slopes at SC and OC via linear fitting on several samples (see Appendix B), while for Saleem is because it has to identify two additional operating points in the curve apart from the inputs SC, OC, and MPP. The execution time is not affected by the PV technology or operating conditions.

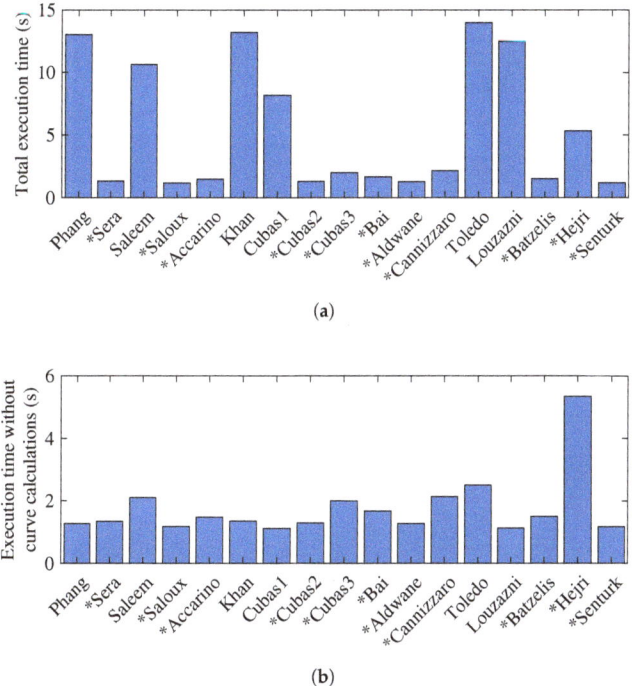

Figure 6. Computational cost of the 17 methods. (**a**) Total execution time. (**b**) Execution time excluding curve calculations.

If the curve calculations cost is excluded (core time), Table 3 and Figure 6b show that the remaining time is much lower in the range of 1–2 s, the same as for the only-datasheet methods. This is because the non-iterative methods involve very few and simple equations, whose evaluation cost is comparably less than the curve calculations. If the additional curve-related data are provided as inputs, calculated beforehand, the last column in Table 3 should be used when assessing the computational performance.

In any case, the times recorded for all methods (with and without the curve calculations cost) are as low as a few microseconds per curve, which is several orders of magnitude less than the numerical or optimization counterparts. As an indication of this huge difference, the iterative method given in [34] was evaluated on the same 1-million dataset using a 20-machine computer cluster, whereas here a common PC was able to complete this task in only a few seconds collectively for all methods.

4.2. The Methods that Rely Only on Datasheet Information

Relying only on datasheet information is commonly considered a desirable feature for a parameter extraction method [19,27,36]; the input data are usually the voltage and current at the SC, OC, MPP and possibly the temperature coefficients (see Table 1). This information is readily available at STC and there is no need for additional measurements, which renders the respective methods more easily and universally applicable. In this section, these techniques are reexamined separately to facilitate selection when there are limitations on the input data.

Table 5 shows the performance of these methods; the main conclusions derived in the previous section do not change, but the best candidates in some assessment criteria are different. In terms of accuracy, the Cannizzaro technique is now the most preferable, yielding the lowest both mean and max errors (along with Cubas3 for the max absolute RMSE). It is worth noting that the four methods that never exceed the 1 A (20%) RMSE (Accarino, Cubas3, Cannizzaro, and Batzelis) are all included in this table, which indicates that the entire I-V curve is not mandatory to achieve good worst-case accuracy. As for the robustness, the methods that yield irregular results in less than 0.1% of the times (Saloux, Cannizzaro, and Batzelis) and the ones that do not fail even once (Saloux, Batzelis, and Senturk) are all of the only-datasheet type (Table 5). Finally, the execution time of all these methods is pretty much the same (except for Hejri as explained in Section 4.1.3), the small differences being attributed to the number of calculation steps. It is worth noting that the methods that employ the Lambert W function (Accarino, Cubas3, Cannizzaro, and Batzelis) do not exhibit any noticeable overhead in the computational time, due to the approximation formulae used for the principal and lower branches (see Appendix A).

In conclusion, these results show that the most favorable options in terms of worst-case accuracy, irregularities, failures, and execution time are methods that do not require additional measurements apart from the ones available in the PV module datasheet. However, when the average accuracy is the main assessment criterion, the Toledo method that needs extra information, yields significantly lower errors.

Table 5. Performance of the methods that rely solely on datasheet information.

Method	Accuracy (RMSE)				Robustness		Complexity
	Absolute (A)		Normalized (%)				
	Mean	Max	Mean	Max	Irregularities	Failures	Total Time (s)
* Sera	0.026	2.471	0.74	49.6	794,670	52	1.3
* Saloux	0.029	1.271	0.79	25.5	2	0	1.2
* Accarino	0.034	0.880	1.09	17.7	2804	25	1.5
* Cubas2	0.026	2.433	0.64	48.8	626,150	53	1.3
* Cubas3	0.024	**0.430**	0.83	13.0	1878	224	2.0
* Bai	0.032	2.544	0.93	51.1	6390	37	1.7
* Aldwane	0.021	1.032	0.59	20.7	416,650	29	1.3
* Cannizzaro	**0.013**	0.431	**0.39**	**11.3**	100	28	2.1
* Batzelis	0.028	0.932	0.87	18.7	412	0	1.5
* Hejri	0.231	6.761	7.21	129.4	794,670	14,057	5.4
* Senturk	0.034	1.285	1.09	25.8	27,847	0	1.2

The asterisk "*" denotes methods that rely only on datasheet information. Bold font indicates the lowest value in the respective column.

4.3. The Methods that Require the SC Slope as Input Data

The extraction of the slope at SC and OC is a challenging task, especially when the I-V curve contains measurement noise or exhibits other types of distortions. Erroneous identification of the slopes has a critical impact on the methods that rely on this data; a factor that affects this identification is the *fitting range*, i.e., the part of the curve used for linear fitting to calculate the slope (see Appendix B). To investigate this dependence, the performance of Phang, Khan, Cubas1, Toledo and Louazni are shown in Table 6 for two different SC slope fitting ranges: 0–10% of V_{oc} (short range—used in the previous sections) and 0–50% of V_{oc} (long range). The former option is theoretically closer to the definition of the SC slope (samples on the SC region) and thus it is used throughout the paper, while the latter is less prone to local distortions and irregularities in the I-V curve.

Apparently, the results in all six performance indicators do not fundamentally differ between the two ranges. No clear conclusion can be extracted on the accuracy, each method being affected differently by the larger range, either positively or negatively, without any noteworthy deviations. Exception to this observation is Toledo, which seems to slightly gain from the larger range in terms of average (absolute and normalized) accuracy, but at the cost of much higher absolute RMSE, although the normalized RMSE is reduced. On the other hand, a clear trend is evident in the robustness: all methods exhibit less irregularities and failures at the larger range, which validates the claim that a larger slope extraction range entails better robustness. It is worth noting that Toledo fails the lowest number of times among the slope-relying methods, which is comparable to some of the only-datasheet counterparts (see also Table 5). The execution times recorded are slightly higher in the larger range case, as expected, due to the higher number of samples to be processed. In fact, the conclusion of this investigation is that a larger fitting range for the extraction of the SC slope handles better distorted I-V curves but does not address the problem and may affect the accuracy.

Table 6. Performance of the methods that require the SC slope as input data.

Method	Fitting Range	Accuracy (RMSE)				Robustness		Complexity	
		Absolute (A)		Normalized (%)				Execution Time (s)	
		Mean	Max	Mean	Max	Irregularities	Failures	Total	Core
Phang	0–10%	0.016	4.210	0.38	69.4	387,440	9952	13.1	1.3
	0–50%	0.013	4.226	0.31	69.7	292,530	9613	15.1	1.4
Khan	0–10%	0.043	4.201	1.19	69.3	511,210	8242	13.2	1.4
	0–50%	0.062	4.199	1.70	69.2	490,640	8207	14.4	1.4
Cubas1	0–10%	0.011	2.445	0.36	49.1	529,530	2445	**8.2**	**1.1**
	0–50%	0.013	2.074	0.32	41.7	330,470	1591	9.3	1.2
Toledo	0–10%	0.007	**1.119**	0.20	97.7	451,220	**392**	14.0	2.5
	0–50%	**0.006**	2.528	**0.17**	51.2	315,110	456	16.4	2.8
Louzazni	0–10%	0.122	1.449	3.43	**27.0**	64,194	30,612	12.5	**1.1**
	0–50%	0.111	1.508	3.18	43.9	**33,147**	25,429	14.2	1.3

Bold font indicates the lowest value in the respective column.

5. Conclusions

This paper performs a rigorous review and assessment of the available non-iterative methods to extract the single-diode PV model parameters. A total of 17 methods are reviewed, implemented, and compared on a dataset of about 1 million I-V curves of six different PV technologies. Some of the main conclusion are:

- Although the simple single-diode model is not the best option for thin-film technologies and low-irradiance measurements, some of the methods perform universally quite decently
- The non-iterative methods are one order of magnitude less accurate than the state-of-the-art numerical counterparts, but require less input data and are much more computationally efficient
- Toledo exhibits the best average accuracy in all PV technologies, whereas the highest worst-case accuracy is brought by a different method in each module type
- Several methods yield negative parameter values (and complex in Hejri) in almost half of the scenarios; this may be acceptable for some applications
- The fewer irregularities are found in Saloux, Cannizzaro, and Batzelis
- Some methods suffer from execution failures quite often, mainly in the HIT and CIGS technologies
- Successful execution is always guaranteed by Accarino, Batzelis, and Senturk
- A major reason for failure is the SC slope extraction due to distorted I-V curve
- A large fitting range for the slope extraction improves slightly the robustness of the slope-relying methods, but has an unpredictable impact on the accuracy
- The calculation times are in the range of microseconds per curve, the higher times observed in the methods that perform calculations on the I-V curve measurements, mainly slope extraction; if this overhead is excluded, the execution time of most methods is quite similar
- The methods that rely only on datasheet information provide the best universal performance in all aspects apart from the average accuracy
- The Lambert W function is related to excellent performance (Accarino, Cubas3, Cannizzaro and Batzelis)

To facilitate selection of the most appropriate method for universal application to all PV technologies, Table 7 below shows the best candidates according to six metrics and accounting for different constraints. Toledo should be selected if focusing on average accuracy and there is access to additional information apart from the datasheet, otherwise the Cannizaro method is more appropriate. When irregular parameters are acceptable to some extent (less than 0.1% of the cases), Cannizzaro yields the best accuracy in terms of both mean and max errors. However, if the execution failures are not acceptable, the Batzelis alternative should be selected instead. Finally, Saloux exhibits always the fewest parameter irregularities and the least execution time, but at the cost of moderate accuracy.

Table 7. Methods with the best universal performance regarding constraints.

Metric	Constraints			
	None	Only-Datasheet	<0.1% Irregularities	No Failures
Mean absolute RMSE	Toledo	Cannizzaro	Cannizzaro	Batzelis
Max absolute RMSE	Cannizzaro, Cubas3	Cannizzaro, Cubas3	Cannizzaro	Batzelis
Mean normalized RMSE	Toledo	Cannizzaro	Cannizzaro	Saloux
Max normalized RMSE	Cannizzaro	Cannizzaro	Cannizzaro	Batzelis
Irregularities	Saloux	Saloux	Saloux	Saloux
Total time	Saloux	Saloux	Saloux	Saloux

Funding: This research has received funding from the European Union's Horizon 2020 research and innovation programme under the Marie Sklodowska-Curie grant agreement No 746638.

Acknowledgments: The author would like to thank Bill Marion and the NREL for their kind provision of the dataset used in the comparative assessment.

Conflicts of Interest: The author declares no conflict of interest.

Appendix A. Computation of the Lambert W Function

The Lambert W function $W\{x\}$ is the inverse of the function we^w, i.e., the root of the equation $we^w = x$ (e.g., $we^w = 2 \leftrightarrow w = W\{2\} = 0.8526$). In general, this relation has several branches in the complex plane, thus not being a function in the conventional sense. However, if the argument x takes real values, then $W\{x\}$ has two real-valued branches: the *principal* branch $W_0\{x\}$ defined for $w \geq -1$ and the *lower* branch $W_{-1}\{x\}$ for $w \leq -1$ [65].

Since the Lambert W function cannot be expressed in terms of other elementary functions, these branches must be calculated either numerically or through approximation formulae. The built-in MATLAB function *lambertw* corresponds to the first approach; one can write *lambertw*(0, x) for $W_0\{x\}$ or *lambertw*(−1, x) for $W_{-1}\{x\}$ to achieve machine accuracy but at an increased computational cost. Alternatively, series expansions are available in the literature, which are applicable only to specific ranges and exhibit some approximation error but are faster and simpler than the numerical approach.

An approximation formula for the principal branch that is applicable to all real positive values is given in [66], which is suitable for evaluation of (2) and (3). However, for the parameter extraction methods presented in Section 3, an even simpler formula may be used [25]:

$$W_0\{x\} = \ln x \left[1 - \frac{\ln(\ln x)}{\ln x + 1}\right]. \tag{A1}$$

This expression is quite accurate for the argument range of Accarino's (29) and Batzelis's (67), thus it is employed here in the implementation of these methods. It should *not* be used, however, for the evaluation of the PV Equations (2) or (3), as it is not applicable to small argument values close to zero.

For the lower branch, [22] adopts an expression proposed in [67]:

$$W_{-1}\{x\} = -1 - \sigma - \frac{2}{M_1}\left(1 - \frac{1}{1 + \frac{M_1\sqrt{\sigma/2}}{1 + M_2\sigma\exp(M_3\sqrt{\sigma})}}\right), \quad \text{(A2)}$$

where $\sigma = -1 - \ln(-x)$, $M_1 = 0.3361$, $M_2 = -0.0042$ and $M_3 = -0.0201$. This equation is used in Cannizzaro's (54) and Cubas3's (44) implementation.

Appendix B. Extraction of the Experimental Resistances (Slopes at SC and OC)

Several parameter extraction methods require as input data the experimental resistances R_{sho} and R_{so}, or equivalently the slope of the *I-V* curve at SC and OC (see (9) in Section 2) [14,18,19,23,24]. Since this information is not available in the PV module datasheet, it can be only extracted through a measured *I-V* curve. Essentially, we need to evaluate the slope of the tangent lines at SC and OC shown in Figure 2 in Section 2.

This process heavily depends on how the *I-V* curve is measured, and the technique used to identify the slope. Usually the *I-V* characteristic is recorded to several tens up to a few hundreds of samples, but this is not standard and depends on the measurement equipment. For the NREL dataset examined in this paper, most curves consist of 180–200 samples. Theoretically, taking the discrete derivative of the two closest samples at the particular point of interest, e.g., SC, should suffice for extracting the slope. However, in practice the *I-V* samples contain measurement noise, they are not that close to each other (differing by 1 V or more at SC), and the curve sometimes features distortions for various reasons, resulting quite often in erroneous slope approximation with this simple approach.

It is generally considered preferable to use a wider part of the characteristic to extract the slope, here referring to as *fitting range*, such as from 0 to 10%, 20% or even 50% of V_{oc} for the SC slope. One can either perform curve-fitting on all the samples included in this range or select two specific points to take the discrete derivative (e.g., SC and MPP). The first approach is adopted in this paper, as it is more robust under any type of *I-V* curve distortions; this is *linear* least squares minimization that involves explicit and computationally efficient formulae (in contrast to the general *non-linear* least squares that is solved iteratively) [68]:

$$\frac{dV}{dI} = \frac{S_{00}S_{20} - S_{10}^2}{S_{00}S_{11} - S_{10}S_{01}}, \quad \text{(A3)}$$

where the auxiliary terms $S_{xy} = \sum_i V_i^x I_i^y$ correspond to sums of sample voltages and currents. For the SC slope, one has to evaluate (A3) using the samples close to SC included in the respective fitting range, e.g., $0 \leq V_i \leq 0.1V_{oc}$, whereas for the OC slope the samples close to OC, e.g., $0 \leq I_i \leq 0.1I_{sc}$. Indicatively, theses ranges include ~15-20 and ~5 samples respectively for the NREL *I-V* curves. This approach is adopted in this paper for all methods that require the SC/OC slopes as input data [14,18,19,23,24], while the formulation of (A3) makes it suitable for embedded applications to microcontrollers. A sensitivity analysis on the fitting range of the SC slope is given in Section 4.3.

References

1. Jordehi, A.R. Parameter estimation of solar photovoltaic (PV) cells: A review. *Renew. Sustain. Energy Rev.* **2016**, *61*, 354–371. [CrossRef]
2. De Soto, W.; Klein, S.; Beckman, W. Improvement and validation of a model for photovoltaic array performance. *Sol. Energy* **2006**, *80*, 78–88. [CrossRef]
3. Laudani, A.; Riganti Fulginei, F.; Salvini, A. Identification of the one-diode model for photovoltaic modules from datasheet values. *Sol. Energy* **2014**, *108*, 432–446. [CrossRef]
4. Laudani, A.; Mancilla-David, F.; Riganti-Fulginei, F.; Salvini, A. Reduced-form of the photovoltaic five-parameter model for efficient computation of parameters. *Sol. Energy* **2013**, *97*, 122–127. [CrossRef]

5. ALQahtani, A.H. A simplified and accurate photovoltaic module parameters extraction approach using matlab. In Proceedings of the 2012 IEEE International Symposium on Industrial Electronics, Hangzhou, China, 28–31 May 2012; pp. 1748–1753.
6. Lineykin, S.; Averbukh, M.; Kuperman, A. An improved approach to extract the single-diode equivalent circuit parameters of a photovoltaic cell/panel. *Renew. Sustain. Energy Rev.* **2014**, *30*, 282–289. [CrossRef]
7. Villalva, M.; Gazoli, J.; Filho, E. Comprehensive approach to modeling and simulation of photovoltaic arrays. *IEEE Trans. Power Electron.* **2009**, *24*, 1198–1208. [CrossRef]
8. Chenni, R.; Makhlouf, M.; Kerbache, T.; Bouzid, A. A detailed modeling method for photovoltaic cells. *Energy* **2007**, *32*, 1724–1730. [CrossRef]
9. Carrero, C.; Ramírez, D.; Rodríguez, J.; Platero, C. Accurate and fast convergence method for parameter estimation of PV generators based on three main points of the I–V curve. *Renew. Energy* **2011**, *36*, 2972–2977. [CrossRef]
10. Mahmoud, Y.A.; Xiao, W.; Zeineldin, H.H. A parameterization approach for enhancing PV model accuracy. *IEEE Trans. Ind. Electron.* **2013**, *60*, 5708–5716. [CrossRef]
11. Cannizzaro, S.; Di Piazza, M.C.; Luna, M.; Vitale, G. Generalized classification of PV modules by simplified single-diode models. In Proceedings of the 2014 IEEE 23rd International Symposium on Industrial Electronics (ISIE), Istanbul, Turkey, 1–4 June 2014; pp. 2266–2273.
12. Accarino, J.; Petrone, G.; Ramos-Paja, C.A.; Spagnuolo, G. Symbolic algebra for the calculation of the series and parallel resistances in PV module model. In Proceedings of the 2013 International Conference on Clean Electrical Power (ICCEP), Alghero, Italy, 11–13 June 2013; pp. 62–66.
13. Nassar-eddine, I.; Obbadi, A.; Errami, Y.; El fajri, A.; Agunaou, M. Parameter estimation of photovoltaic modules using iterative method and the Lambert W function: A comparative study. *Energy Convers. Manag.* **2016**, *119*, 37–48. [CrossRef]
14. Phang, J.C.H.; Chan, D.S.H.; Phillips, J.R. Accurate analytical method for the extraction of solar cell model paramaters. *Electron. Lett.* **1984**, *20*, 406–408. [CrossRef]
15. Sera, D.; Teodorescu, R.; Rodriguez, P. Photovoltaic module diagnostics by series resistance monitoring and temperature and rated power estimation. In Proceedings of the 2008 34th Annual Conference of IEEE Industrial Electronics, Orlando, FL, USA, 10–13 November 2008; pp. 2195–2199.
16. Saleem, H.; Karmalkar, S. An analytical method to extract the physical parameters of a solar cell from four points on the illuminated J-V curve. *IEEE Electron Device Lett.* **2009**, *30*, 349–352. [CrossRef]
17. Saloux, E.; Teyssedou, A.; Sorin, M. Explicit model of photovoltaic panels to determine voltages and currents at the maximum power point. *Sol. Energy* **2011**, *85*, 713–722. [CrossRef]
18. Khan, F.; Baek, S.H.; Park, Y.; Kim, J.H. Extraction of diode parameters of silicon solar cells under high illumination conditions. *Energy Convers. Manag.* **2013**, *76*, 421–429. [CrossRef]
19. Cubas, J.; Pindado, S.; Victoria, M. On the analytical approach for modeling photovoltaic systems behavior. *J. Power Sources* **2014**, *247*, 467–474. [CrossRef]
20. Bai, J.; Liu, S.; Hao, Y.; Zhang, Z.; Jiang, M.; Zhang, Y. Development of a new compound method to extract the five parameters of PV modules. *Energy Convers. Manag.* **2014**, *79*, 294–303. [CrossRef]
21. Aldwane, B. Modeling, simulation and parameters estimation for Photovoltaic module. In Proceedings of the 2014 1st International Conference on Green Energy ICGE 2014, Sfax, Tunisia, 25–27 March 2014; pp. 101–106.
22. Cannizzaro, S.; Di Piazza, M.C.; Luna, M.; Vitale, G. PVID: An interactive Matlab application for parameter identification of complete and simplified single-diode PV models. In Proceedings of the 2014 IEEE 15th Workshop on Control and Modeling for Power Electronics (COMPEL), Santander, Spain, 22–25 June 2014.
23. Toledo, F.; Blanes, J.M. Geometric properties of the single-diode photovoltaic model and a new very simple method for parameters extraction. *Renew. Energy* **2014**, *72*, 125–133. [CrossRef]
24. Louzazni, M.; Aroudam, E.H. An analytical mathematical modeling to extract the parameters of solar cell from implicit equation to explicit form. *Appl. Sol. Energy* **2015**, *51*, 165–171. [CrossRef]
25. Batzelis, E.I.; Papathanassiou, S.A. A method for the analytical extraction of the single-diode PV model parameters. *IEEE Trans. Sustain. Energy* **2016**, *7*, 504–512. [CrossRef]
26. Hejri, M.; Mokhtari, H.; Azizian, M.R.; Söder, L. An analytical-numerical approach for parameter determination of a five-parameter single-diode model of photovoltaic cells and modules. *Int. J. Sustain. Energy* **2016**, *35*, 396–410. [CrossRef]

27. Senturk, A.; Eke, R. A new method to simulate photovoltaic performance of crystalline silicon photovoltaic modules based on datasheet values. *Renew. Energy* **2017**, *103*, 58–69. [CrossRef]
28. Murtaza, A.; Munir, U.; Chiaberge, M.; Di Leo, P.; Spertino, F. Variable Parameters for a Single Exponential Model of Photovoltaic Modules in Crystalline-Silicon. *Energies* **2018**, *11*, 2138. [CrossRef]
29. Kumar, M.; Kumar, A. An efficient parameters extraction technique of photovoltaic models for performance assessment. *Sol. Energy* **2017**, *158*, 192–206. [CrossRef]
30. Ibrahim, H.; Anani, N. Evaluation of Analytical Methods for Parameter Extraction of PV modules. *Energy Procedia* **2017**, *134*, 69–78. [CrossRef]
31. Xiong, G.; Zhang, J.; Yuan, X.; Shi, D.; He, Y.; Yao, G. Parameter extraction of solar photovoltaic models by means of a hybrid differential evolution with whale optimization algorithm. *Sol. Energy* **2018**, *176*, 742–761. [CrossRef]
32. Soon, J.J.; Low, K.S. Photovoltaic model identification using particle swarm optimization with inverse barrier constraint. *IEEE Trans. Power Electron.* **2012**, *27*, 3975–3983. [CrossRef]
33. Askarzadeh, A.; Dos Santos Coelho, L. Determination of photovoltaic modules parameters at different operating conditions using a novel bird mating optimizer approach. *Energy Convers. Manag.* **2015**, *89*, 608–614. [CrossRef]
34. Toledo, F.J.; Blanes, J.M.; Galiano, V. Two-Step Linear Least-Squares Method For Photovoltaic Single-Diode Model Parameters Extraction. *IEEE Trans. Ind. Electron.* **2018**, *65*, 6301–6308. [CrossRef]
35. Lim, L.H.I.; Ye, Z.; Ye, J.; Yang, D.; Du, H. A linear identification of diode models from single I-V characteristics of PV panels. *IEEE Trans. Ind. Electron.* **2015**, *62*, 4181–4193. [CrossRef]
36. Ciulla, G.; Lo Brano, V.; Di Dio, V.; Cipriani, G. A comparison of different one-diode models for the representation of I–V characteristic of a PV cell. *Renew. Sustain. Energy Rev.* **2014**, *32*, 684–696. [CrossRef]
37. Chin, V.J.; Salam, Z.; Ishaque, K. Cell modelling and model parameters estimation techniques for photovoltaic simulator application: A review. *Appl. Energy* **2015**, *154*, 500–519. [CrossRef]
38. Humada, A.M.; Hojabri, M.; Mekhilef, S.; Hamada, H.M. Solar cell parameters extraction based on single and double-diode models: A review. *Renew. Sustain. Energy Rev.* **2016**, *56*, 494–509. [CrossRef]
39. Boutana, N.; Mellit, A.; Lughi, V.; Massi Pavan, A. Assessment of implicit and explicit models for different photovoltaic modules technologies. *Energy* **2017**, *122*, 128–143. [CrossRef]
40. Piazza, M.C.D.; Luna, M.; Petrone, G.; Spagnuolo, G. Translation of the Single-Diode PV Model Parameters Identified by Using Explicit Formulas. *IEEE J. Photovolt.* **2017**, *7*, 1009–1016. [CrossRef]
41. Peñaranda Chenche, L.E.; Hernandez Mendoza, O.S.; Bandarra Filho, E.P. Comparison of four methods for parameter estimation of mono- and multi-junction photovoltaic devices using experimental data. *Renew. Sustain. Energy Rev.* **2018**, *81*, 2823–2838. [CrossRef]
42. Marion, B.; Anderberg, A.; Deline, C.; del Cueto, J.; Muller, M.; Perrin, G.; Rodriguez, J.; Rummel, S.; Silverman, T.J.; Vignola, F.; et al. New data set for validating PV module performance models. In Proceedings of the 2014 IEEE 40th Photovoltaic Specialist Conference (PVSC), Denver, CO, USA, 8–13 June 2014; pp. 1362–1366.
43. Batzelis, E.I. Simple PV performance equations theoretically well-founded on the single-diode model. *IEEE J. Photovolt.* **2017**, *7*, 1400–1409. [CrossRef]
44. Karatepe, E.; Boztepe, M.; Çolak, M. Development of a suitable model for characterizing photovoltaic arrays with shaded solar cells. *Sol. Energy* **2007**, *81*, 977–992. [CrossRef]
45. Quaschning, V.; Hanitsch, R. Numerical simulation of current-voltage characteristics of photovoltaic systems with shaded solar cells. *Sol. Energy* **1996**, *56*, 513–520. [CrossRef]
46. Pandey, P.K.; Sandhu, K. Multi diode modelling of PV cell. In Proceedings of the 2014 IEEE 6th India International Conference on Power Electronics (IICPE), Kurukshetra, India, 8–10 December 2014.
47. Kawamura, H.; Naka, K.; Yonekura, N.; Yamanaka, S.; Kawamura, H.; Ohno, H.; Naito, K. Simulation of I–V characteristics of a PV module with shaded PV cells. *Sol. Energy Mater. Sol. Cells* **2003**, *75*, 613–621. [CrossRef]
48. Merten, J.; Asensi, J.; Voz, C.; Shah, A.; Platz, R.; Andreu, J. Improved equivalent circuit and analytical model for amorphous silicon solar cells and modules. *IEEE Trans. Electron Devices* **1998**, *45*, 423–429. [CrossRef]

49. Abbassi, A.; Gammoudi, R.; Ali Dami, M.; Hasnaoui, O.; Jemli, M. An improved single-diode model parameters extraction at different operating conditions with a view to modeling a photovoltaic generator: A comparative study. *Sol. Energy* **2017**, *155*, 478–489. [CrossRef]
50. Celik, A.N.; Acikgoz, N. Modelling and experimental verification of the operating current of mono-crystalline photovoltaic modules using four- and five-parameter models. *Appl. Energy* **2007**, *84*, 1–15. [CrossRef]
51. Hejri, M.; Mokhtari, H.; Azizian, M.R.; Ghandhari, M.; Soder, L. On the parameter extraction of a five-parameter double-diode model of photovoltaic cells and modules. *IEEE J. Photovolt.* **2014**, *4*, 915–923. [CrossRef]
52. Chouder, A.; Silvestre, S.; Sadaoui, N.; Rahmani, L. Modeling and simulation of a grid connected PV system based on the evaluation of main PV module parameters. *Simul. Model. Pract. Theory* **2012**, *20*, 46–58. [CrossRef]
53. Kou, Q.; Klein, S.; Beckman, W. A method for estimating the long-term performance of direct-coupled PV pumping systems. *Sol. Energy* **1998**, *64*, 33–40. [CrossRef]
54. Tian, H.; Mancilla-David, F.; Ellis, K.; Muljadi, E.; Jenkins, P. A cell-to-module-to-array detailed model for photovoltaic panels. *Sol. Energy* **2012**, *86*, 2695–2706. [CrossRef]
55. Khezzar, R.; Zereg, M.; Khezzar, A. Modeling improvement of the four parameter model for photovoltaic modules. *Sol. Energy* **2014**, *110*, 452–462. [CrossRef]
56. Cubas, J.; Pindado, S.; de Manuel, C. Explicit expressions for solar panel equivalent circuit parameters based on analytical formulation and the Lambert W-function. *Energies* **2014**, *7*, 4098–4115. [CrossRef]
57. Petrone, G.; Ramos-Paja, C.A.; Spagnuolo, G. *Photovoltaic Sources Modeling*, 1st ed.; Wiley-IEEE Press: Hoboken, NJ, USA, 2017; p. 208.
58. Mahmoud, Y.; El-Saadany, E.F. Fast power-peaks estimator for partially shaded PV systems. *IEEE Trans. Energy Convers.* **2016**, *31*, 206–217. [CrossRef]
59. Anderson, A.J. *Photovoltaic Translation Equations: A New Approach*; National Renewable Energy Laboratory: Golden, CO, USA, 1996.
60. Chan, D.S.H.; Phillips, J.R.; Phang, J.C.H. A comparative study of extraction methods for solar cell model parameters. *Solid State Electron.* **1986**, *29*, 329–337. [CrossRef]
61. Chan, D.S.H.; Phang, J.C.H. Analytical methods for the extraction of solar-cell single- and double-diode model parameters from I-V characteristics. *IEEE Trans. Electron Devices* **1987**, *ED*, 286–293. [CrossRef]
62. Hadj Arab, A.; Chenlo, F.; Benghanem, M. Loss-of-load probability of photovoltaic water pumping systems. *Sol. Energy* **2004**, *76*, 713–723. [CrossRef]
63. Orioli, A.; Di Gangi, A. A procedure to calculate the five-parameter model of crystalline silicon photovoltaic modules on the basis of the tabular performance data. *Appl. Energy* **2013**, *102*, 1160–1177. [CrossRef]
64. Technical Guideline. *Generic Solar Photovoltaic System Dynamic Simulation Model Specification*; Sandia National Laboratorie: Livermore, CA, USA, 2013.
65. Corless, R.M.; Gonnet, G.H.; Hare, D.E.G.; Jeffrey, D.J.; Knuth, D.E. On the Lambert W function. *Adv. Comput. Math.* **1996**, *5*, 329–359. [CrossRef]
66. Batzelis, E.I.; Routsolias, I.A.; Papathanassiou, S.A. An explicit PV string model based on the Lambert W function and simplified MPP expressions for operation under partial shading. *IEEE Trans. Sustain. Energy* **2014**, *5*, 301–312. [CrossRef]
67. Barry, D.; Parlange, J.Y.; Sander, G.; Sivaplan, M. A class of exact solutions for Richards' equation. *J. Hydrol.* **1993**, *142*, 29–46. [CrossRef]
68. Xiao, W.; Lind, M.; Dunford, W.; Capel, A. Real-time identification of optimal operating points in photovoltaic power systems. *IEEE Trans. Ind. Electron.* **2006**, *53*, 1017–1026. [CrossRef]

© 2019 by the author. Licensee MDPI, Basel, Switzerland. This article is an open access article distributed under the terms and conditions of the Creative Commons Attribution (CC BY) license (http://creativecommons.org/licenses/by/4.0/).

Article

Experimental Research of Transmissions on Electric Vehicles' Energy Consumption

Polychronis Spanoudakis [1,*], Nikolaos C. Tsourveloudis [1], Lefteris Doitsidis [2] and Emmanuel S. Karapidakis [3]

1. School of Production Engineering & Management, Technical University of Crete, 73100 Chania, Greece; nikost@dpem.tuc.gr
2. Department of Electronic Engineering, School of Applied Sciences, Technological Educational Institute of Crete, 73133 Chania, Greece; ldoitsidis@chania.teicrete.gr
3. Department of Electrical Engineering, School of Engineering, Technological Educational Institute of Crete, 71410 Heraklion, Greece; karapidakis@staff.teicrete.gr
* Correspondence: hroniss@dpem.tuc.gr; Tel.: +30-282-103-7427

Received: 10 December 2018; Accepted: 22 January 2019; Published: 26 January 2019

Abstract: The growth of electric vehicles share of total passenger-vehicle sales is evident and is expected to be a very big market segment by 2030. Range of travel and pricing are the most influencing factors that affect their gain in market share. As so, powertrain development is a key technology factor researched by the automotive industry. To explore, among others, how the energy consumption of zero emission vehicles is affected by different transmissions, we developed, built and installed a variety of them on a custom hydrogen fuel cell powered urban vehicle. In this work we present a comparison of the effect, on the energy consumption of the proposed testbed, of a prototype custom build 2-speed gearbox and a single stage transmission. Results presented show a reduction of the overall energy consumption with the use of the 2-speed gearbox, compared to single stage, as well as the effect of gear change speed, related to speed, in energy consumption. Finally, a correlation of experimental results using a custom build CVT is conducted compared to single stage transmission. A comparison to simulation results found in literature is performed for all the transmissions tested on road.

Keywords: transmission; electric vehicle; 2-speed; gear change; CVT; energy consumption

1. Introduction

There is an increasing demand for Electric Vehicles (EV). In 2017, 1.3 million units were purchased around the world. The aforementioned number corresponds to 1% of the total passenger-vehicle sales, and corresponds to a 57% increase from 2016 numbers [1]. Some of the biggest original equipment manufacturers (OEMs) have announced the introduction of more than 100 original models powered by an electric motor by 2024, while the total share of electric vehicles is estimated to reach 30–35% in major markets, and 20–25% globally by 2030. EVs market share will be boosted by their ability to reach higher ranges and their target to increase design efficiency and reduce manufacturing cost to become affordable to more customer segments. Currently, after extended testing in different types of EVs, the average range is more than 300 Km, and has surpassed the expectation of the largest customer segments, which combined with the current (reduced) price tag in EVs, shows the increased potential of the aforementioned market.

To lower the fuel consumption and increase the autonomy, different types of engines and power transmission components are investigated. Currently the automotive manufacturers are focusing either in new or in optimized powertrain systems that can achieve better fuel consumption. These types of designs are developed not only for internal combustion engines but also for the case of hybrid/electric cars [2].

The majority of production vehicles use conventional Manual Transmissions (MTs) or Automatic Transmissions (ATs), while a viable alternative are the Continuously Variable Transmissions (CVTs) and infinitely variable transmissions. A detailed study of their characteristics, including but not limited to their efficiency, torque transfer and applicability is presented in [3–5].

The torque characteristic is the main advantage of an electric motor, since it can provide the maximum value of the available torque from completely stopped up to relatively low speeds, and then it is proportional to the maximum power as the motor's speed is increasing [6]. To exploit the aforementioned advantage, most of the commercially available EVs are using powertrains which are directly connected to the driving wheels via a single reduction ratio, having several competitive advantages [3].

However, gearbox use can be a solution in order to meet the design requirements of vehicle acceleration and maximum speed, where high-performance traction motor and batteries are required. Thus, a 2-speed or even an n-speed gearbox may be installed to increase the available wheel torque, to reduce acceleration times and increase the achievable road grade, whilst prompting the power source to operate in a higher efficiency region during drive cycles. A multiple-speed transmission adopted for an electric drivetrain has energy consumption benefits over a single-speed equivalent [7]. There are two main advantages that cope with that. When using a two-speed transmission the first gear ratio can be chosen to increase the low speed torque to improve acceleration and increase the achievable road grade, whereas the second gear ratio can be reduced to extend the operating vehicle speed range. The second major advantage from adopting a multiple speed transmission is that the drivetrain can theoretically operate in a higher efficiency region for a larger portion of a driving cycle [6].

Current state-of-the-art on gearbox or alternative transmissions use on electric vehicles (EVs) is found in [8–12] including a vast number of simulation based comparisons using 2-speed versus single speed gearbox or CVT's use [6–8,13–17]. In [18] different electric drivetrain configurations were discussed along with the implications of installing a multiple speed transmission in a fully electric drivetrain. In [19] a vehicle model was developed consisting of physical models of the components considering the moments of inertia, drag torques and efficiency maps which accounted for the variation on temperature (specifically the electric motor). Furthermore, an optimisation of the gear ratios for both transmissions and shift points for the two-speed was undertaken. The two-speed vehicle evidently operates in the high efficiency region for a larger portion of the driving cycle giving rise to a reduction in energy consumption. Specifically, it was found that on flat ground above 35 km/h, the two-speed is saving energy being in second gear as opposed to first gear so spends the majority of the driving cycle in second gear. The authors also compared the two drivetrains with the same electric motor and found a 5–10% energy consumption improvement in favour of the two-speed system.

In [20] the authors went on to analyse two-speed, three-speed and four-speed drivetrains with gear ratios selected based on the results of the CVT gear ratio optimisation. The results show a marked improvement over the single-speed drivetrain by using a multiple-speed transmission with energy consumption gains ranging from 4.5 % to 11 % over different driving cycles. It is evident that further gains can be made by adopting a CVT, however, the marginal gains over the four-speed transmission would in fact be lost in the additional transmission losses between the two systems. Finally, some concept gearbox prototypes targeting mostly EVs use are already presented in the market, such as those by Vocis/Oerlikon Graziano, Antonov and Kreisler Electric [21–23].

According to the above, there is limited research on real on road tests, in order to identify if gearbox use can provide a feasible solution towards reduced energy consumption, as also to define the effects of gear change. As already mentioned, to the best of our knowledge such research results have been only based in simulation results. In the proposed work we present a comparative analysis of a custom 2-speed gearbox and a flat belt driven CVT, developed by our research team with a single stage transmission. Energy consumption and different gear change speed effects on an urban electric vehicle are presented, based on road tests experimental results and comparisons. These results are also used as an evaluation tool compared to literature simulation results.

The rest of the manuscript is organised as follows, in Section 2 the experimental set-up is presented including the detailed vehicle specifications, the description of the powertrain and the drivetrain as well as the description of the on the road experimental procedure and set-up. In Section 3 we describe in detail the experiments conducted with the different configurations and we provide a comparison between the different configurations. Finally, in Section 4 we conclude with some key findings and discussion for future research.

2. Experimental Set-Up

2.1. Vehicle Specifications

For the evaluation of the proposed system, we conducted tests using realistic conditions. Our testbed was the custom build urban vehicle ER16 (Figure 1). The powertrain is an electric motor which uses as a power source an H_2 fuel cell. The drivetrain (standard transmission) is composed from a single-stage geared transmission with ratio 10:1. The main technical characteristics of the vehicle are presented at Table 1.

Figure 1. The prototype testbed vehicle ER16.

Table 1. Testbed Vehicle Specification.

Characteristic	
Chassis	Aluminum alloy
Body	Carbon fiber
Motor	Brushless electric motor
Max Motor Torque	4 Nm
Max Motor rpm	4000 rpm
Power Source	H_2 fuel cell, 1.2 KW
Dimension	2.5 × 1.25 × 1 m (L × W × H)
Weight	77 Kg/79 Kg (with gearbox)
Max Vehicle Speed	37 Km/h

2.2. Powertrain and Drivetrain Setup

Two different powertrain configurations were tested, (i) a single-stage transmission (Figure 2) and (ii) a 2-speed change gearbox (Figure 3). In both cases the electric motor is powered by a hydrogen fuel cell using H_2 at 200 bar. In the first case the transmission system is positioned between the motor and the wheel so that it can provide the needed torque and rpm, and works as a final transmission (transmission ratio 10:1). In the second case, with the installation of the system a two-stage transmission system occurs, while the rest of the powertrain remains unchanged. The second stage (final transmission) is kept exactly the same providing a 10:1 ratio. The first stage (gearbox) provides a gear change with ratios 1.4:1 and 1:1 respectively. Thus, the total transmission ratios are 14:1 (1st gear) and 10:1 (2nd gear), which are mostly setup and used as launch and cruise vehicle driving modes.

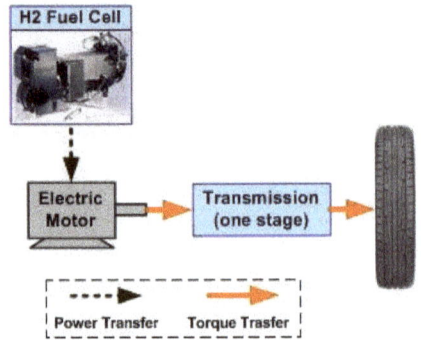

Figure 2. Testbed vehicle standard powertrain configuration.

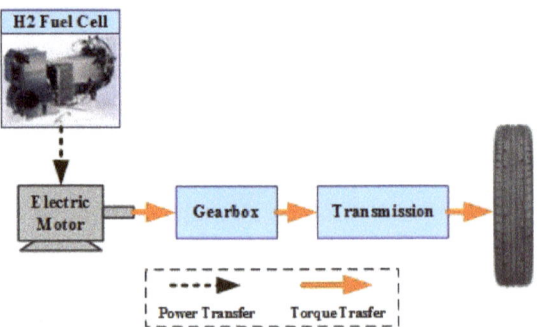

Figure 3. Testbed vehicle powertrain configuration with installed gearbox.

The first stage of the transmission presented is custom built for the configuration and characteristics of the specific testbed vehicle. It is the second version of such a gearbox developed by our research team. The first, introduced in 2014, was integrated in the ER14 vehicle weighing 2.3 Kg and used the exact same gear ratios [24]. The version presented here has been redeveloped, providing higher gear change capabilities, better efficiency and lower weight (Figure 4). A major improvement was in the shifting mechanism providing synchronization capabilities as used in production cars. The synchromesh system consists of two straight-cut gear sets that are constantly "meshed" together and a dog clutch is used for changing gears. Synchro-rings are used in addition to the dog clutch to closely match the rotational speeds of the two sides of the transmission before making a full mechanical engagement. An electronic shifting control is also installed, using a small electric motor to provide linear movement of the dog-clutch to shift between gears, commanded by a switch on driver's steering wheel. The speed of shifting (related to electric motor rpm) was experimentally proven by testing, so an adequate and efficient operation is accomplished in various speeds of operation. The total gearbox weight is measured at 1.9 Kg, using aluminum alloy material (6063T6) for most parts, except the gears, which are made from plastic.

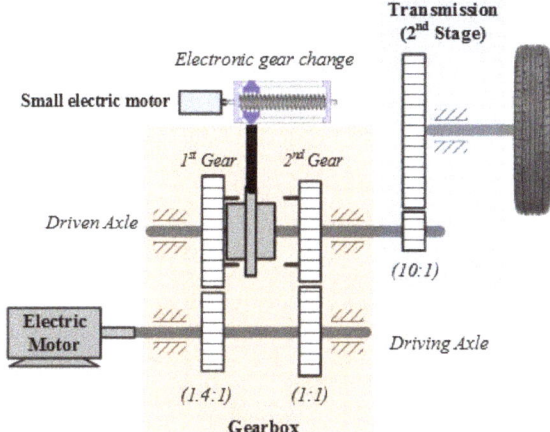

Figure 4. Detailed drawing of 1st and 2nd transmission stages, showing electronically controlled gear change mechanism and corresponding gear ratios.

As mentioned, the use of this gearbox mainly targets lower energy consumption during launch as well as electric motor operation at a higher efficiency region. The construction of the gearbox is based on two gear pairs attached on two parallel axles (Figure 4). The gears are fixed on the first axle (driving axle) and are rotated by the electric motor. On the second axle (driven axle), every gear has a bearing installed that provides a free rotation on the axle. Using the synchronizing mechanism placed between the gears of the second axle, we can change from one gear ratio (1st gear) to another (2nd gear). When no gear is selected, both gears of the second axle can rotate free and thus no power is transferred to the second transmission stage and to the wheel. If the first or second gear is selected by the synchronizing mechanism, then power is transferred to the second stage of the transmission using the corresponding gear ratio. In Figure 5 the custom build gearbox is presented as installed on the prototype vehicle. Using the 2-speed change transmission, the different gear change points occur on different vehicle speeds but for the exactly same gear ratios. Therefore, the main effect is gear change point (speed) contribution, which corresponds to altered motor rpm operation and this leads to a reduced power request.

Figure 5. A view of the custom built prototype 2-speed gearbox.

The efficiency map of the electric motor was not provided by the manufacturer. Therefore, series of experimental testing were conducted in a custom testbed, in order to measure electric motor efficiency at specific loads [12]. In our case, testing targeted vehicle operation at the speed of 25 Km/h, where power demand for the prototype vehicle is 200 W. The custom testbed setup includes: *a rotary torque transducer* connected to the electric motor axle (input) and to a *hydraulic disc brake* (output), as well as amperometer and voltmeter measurements from the motor input. The data recorded are:

1. Output data from the torque transducer (Motor speed (rpm), Torque (Nm) and Power (Kw)).
2. Input data from the power measurements in the input of the electric motor.

In all measurements, external load is set to 200 W by applying pressure to the hydraulic disc brake and data are recorder every 500 rpm. The results of the motor's efficiency are presented in Figure 6. As shown, maximum motor efficiency (91.3%) is reached at 2500 rpm. In addition, an optimal rpm range of motor operation is from 2300 to 2500 rpm, where efficiency exceeds 90%.

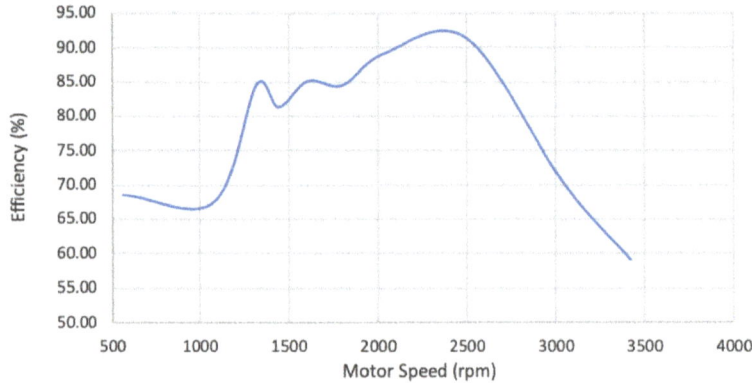

Figure 6. Motor's Efficiency vs. motor speed.

Motor torque measurements versus motor speed are also presented in Figure 7. The maximum torque found is 4 Nm and drops as motor speed raises. Measurements recorded for a motor speed up to 3500 rpm due to testbed limitations for adequate results.

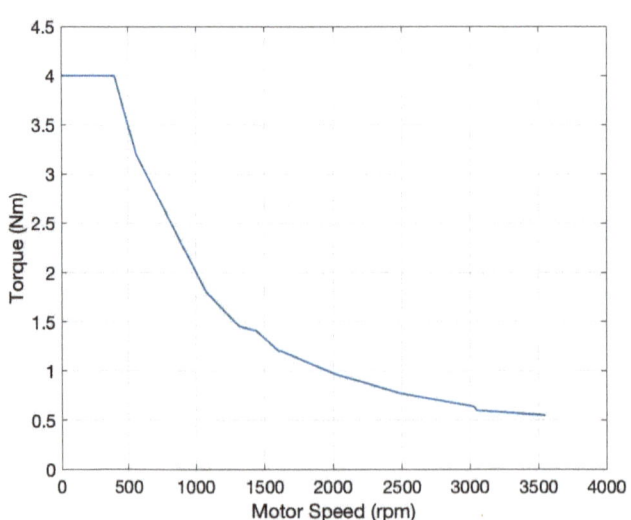

Figure 7. Motor torque measurements vs. motor speed.

Regarding the efficiency of the drivetrain, several measurements and assumptions were made. Specifically, the single-stage transmission, consisting of one gear pair, is assumed to have an efficiency of 95%. This efficiency is also used as final transmission efficiency when the 2-speed gearbox or CVT is installed. For the 2-speed transmission, the efficiency is assumed at 95%. Thus in this case a final drivetrain efficiency of 90% is obtained.

For the CVT, extensive experimental testing has been conducted in order to measure the efficiency on a laboratory testbed [8]. The testbed setup is identical to the one used for motor efficiency measurements. Its main components are: (a) a rotary torque transducer measuring motor speed (rpm), Torque (Nm) and Power (Kw), (b) a hydraulic disc brake to apply external loads, (c) a brake handle, to regulate torque, (d) a brushless electric motor providing the input torque, (e) throttle adjustment for motor speed calibration and (f) a laptop for data recording. Measurements are conducted in respect to electric motor maximum torque and best efficiency speed (2500 rpm). Efficiency ranges to a maximum of 93%. As expected, best efficiency is found at higher motor speeds (2500 rpm). In this case the final drivetrain efficiency is calculated as 88.4%.

2.3. On Road Experimental Setup

All experiments were conducted on TUC's campus, in a track with a total length of 240 m (Figure 8), while a full test consisted of two laps (480 m). Initially we performed experiments with the standard setup and then we replicated them using the 2-speed gearbox, in order to compare their respective energy consumption. To perform the aforementioned comparison, in the second case, different gear change speeds were chosen as shift speeds (8, 11, 14, 17 and 20 Km/h). The vehicle was starting 60 m (Figure 8) before the first corner (1). In order to minimize the driver's intervention and extract unbiased results, the following ruleset was followed:

1. Speed limits are set for the first time the car passes from corners (1) and (2), set at 15–17 Km/h. The driver tries to follow this rule at almost constant acceleration.
2. Above 22 km/h the vehicle increased its velocity using full throttle, which results in higher energy consumption, without actual interference from the driver.
3. Total time to complete the two laps is set to a margin of 79–83 s with a max speed of 30 Km/h. Driver cannot brake to increase lap time but can stop pushing the throttle while cornering in order to adjust his lap timeframe.

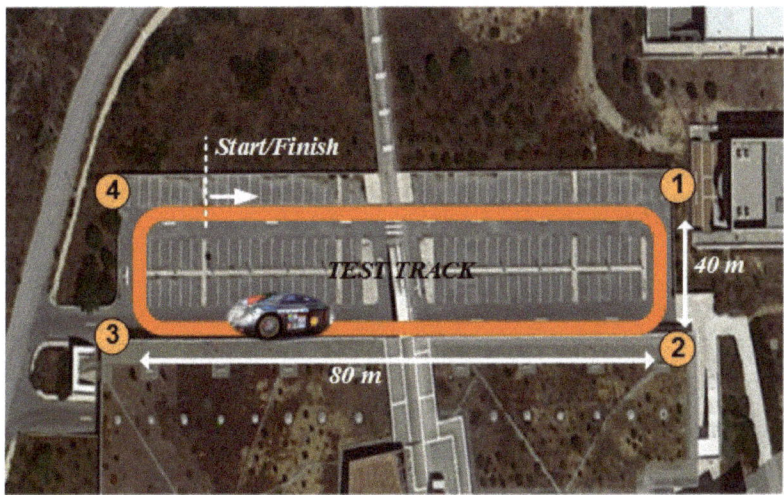

Figure 8. Test track specifications, used for on road tests.

2.4. Data Recording

The acquisition of data streams used for evaluation of tests gathered in road conditions consists of a custom data logger, tailored to serve the needs of the TUCer team. The custom data logger device (TUCer logger) records data of power supply voltage and power supply current, with frequency of a sample set every 0.5 s (2 Hz). Voltage and current are measured at the input of the electric motor. The core of the logger is the popular Arduino Uno microcontroller. Voltage measurement is done by a voltage divider. Current measurement by a shunt resistor and an AD623 Instrumentation Amplifier. All inputs are transferred to the Arduino for processing and evaluation. To address the Arduino's lack of internal memory (for logging purposes), a serial connection (via USB) to a Raspberry Pi is established. The Raspberry Pi receives the data from the Arduino and logs a file for retrieval after the test. For the measurement of hydrogen consumption a flow meter of high accuracy is used (Red-y compact), providing real time measurement of total hydrogen consumption. In every test, measurement of the flow meter is recorded at the start and finish of the two laps. The flow meter is placed at the input of the fuel cell.

3. On Road Testing and Results

Based on the setup and procedures described in detail in Section 2, we measured the power demand (Watt) versus sampling time for different scenarios (Figures 9–14). We also recorded and calculated the mean power and total hydrogen consumption, which is proportional to power demand, following a curve according to Fuel Cell specifications [24]. Three different categories of experiments are presented: Single-stage setup testing, 2-speed gearbox setup testing and CVT setup testing, respectively.

3.1. Tests on Single-Stage Setup

The first scenario concerns the standard powertrain configuration and after conducting experiments as described in detail in Section 2, we calculate the power demand presented in Figure 9. During cornering the power is minimized, where the driver stops pushing the throttle but without braking. On the other parts of the track power demand reaches 400 W. The H_2 consumption was measured at 4.43 lt and mean power demand was calculated at 213.45 W. The total time on track was 83 s and the maximum speed reached was 29.8 Km/h.

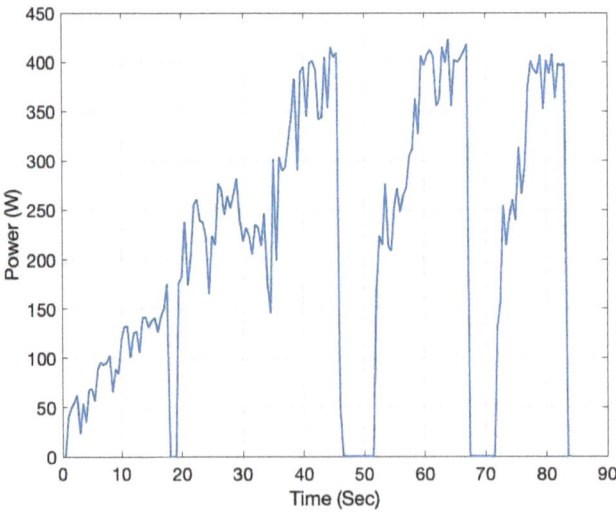

Figure 9. Single-stage setup testing.

3.2. Tests on Two-Speed Gearbox Setup

The second scenario concerns the two-speed gearbox, where the main differentation is the gear shift speed, which has a significant impact on the testbed's performance. For assessment purposes gear change from 14:1 to 10:1 is set at five different vehicle speeds (8, 11, 14, 17 and 20 Km/h) during launch. The scope of this procedure is to asses the impact of gear shift, at different speeds, to energy consumption (Table 2).

Table 2. Impact of gear change at different vehicle speeds.

Gear Change Speed (Km/h)	Mean Power Demand (W)	H_2 (lt)
8	210.12	4.37
11	205.38	4.28
14	221.32	4.49
17	232.09	4.58
20	235.82	4.62

Using the gearbox provides higher acceleration which corresponds to less time to reach the first corner (compared to single-stage) and total time in track is found between 80–81 s for both laps. Maximum speed reached was 29.2–30 Km/h. As so, the vehicle's performance is almost the same in terms of speed inside the track, but it is clear that gearbox use can achieve lower track time. The results are presented in Figures 10–14 whereas the minimum energy consumption is at 11 km/h gear change speed (H_2 consumption: 4.28 lt, mean power demand: 205.38 W), followed by 8 km/h gear change as second best (H_2 consumption: 4.37 lt, mean power demand: 210.12 W). Above 11 km/h gear shift speed, mean power demand is higher. Compared to single stage transmission results (H_2 consumption: 4.43 lt, mean power demand: 213.45 W), correspond to 3.4% H_2 consumption reduction and 3.8% lower mean power demand.

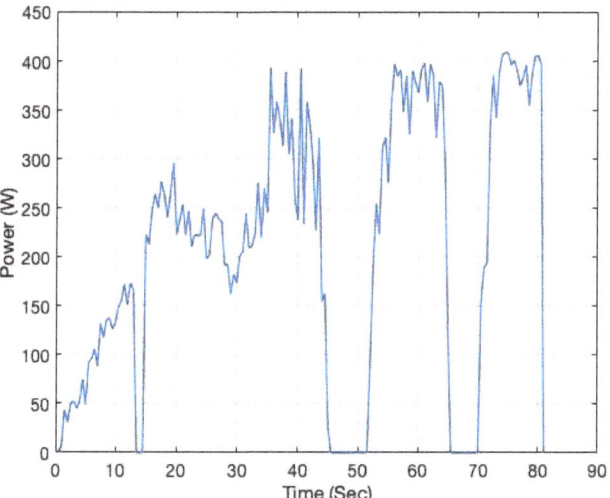

Figure 10. Two-Speed gearbox testing (gear change at 8 km/h).

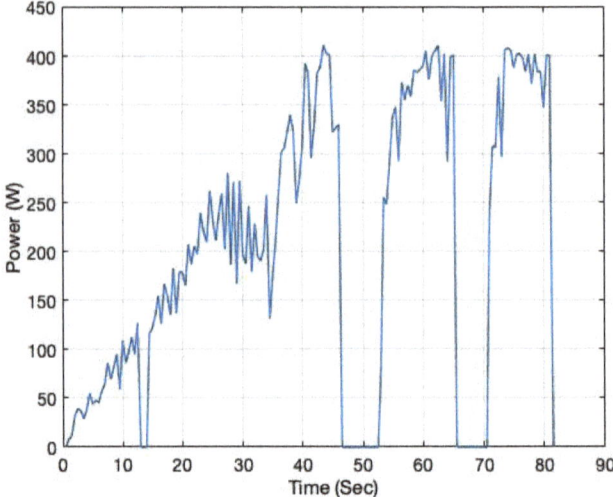

Figure 11. Two-Speed gearbox testing (gear change at 11 km/h).

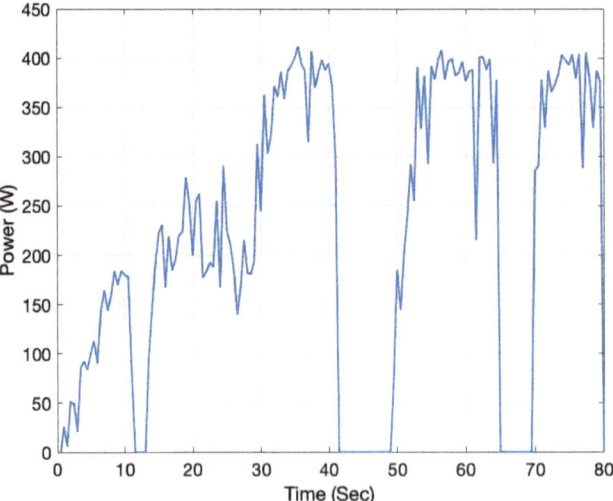

Figure 12. Two-Speed gearbox testing (gear change at 14 km/h).

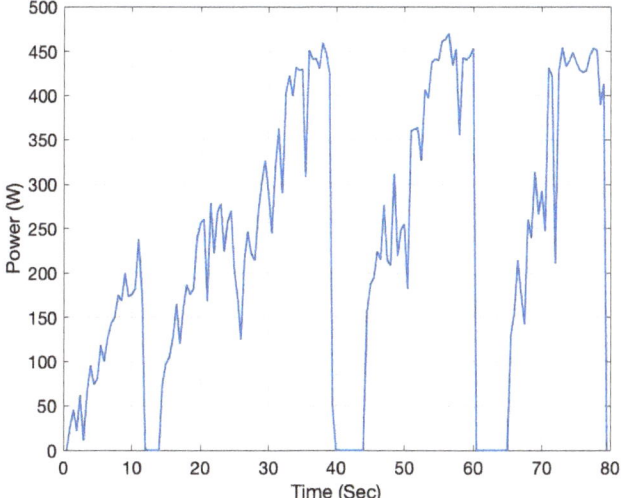

Figure 13. Two-Speed gearbox testing (gear change at 17 km/h).

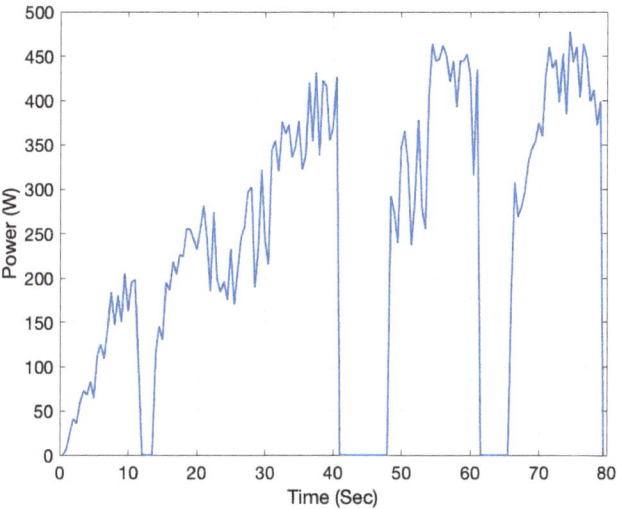

Figure 14. Two-Speed gearbox testing (gear change at 20 km/h).

3.3. On Road Testing of a Prototype CVT

The prototype CVT transmission used was named as an Electronic Shift Variable Transmission (ESVT) and was installed on the ER12 prototype urban vehicle build by TUC Eco Racing team (Technical University of Crete). It is equipped with a real time electronic control of ratio change according to motor rpm, in order to achieve higher motor efficiency and lower energy consumption on a driving cycle. Its operation targets higher traction force thus there is lower consumption at vehicle launch and electronically shifted variable gear ratios ranging from 1.5 to 1 at higher speeds, resulting in electric motor operation in a better efficiency region. In addition to a fixed secondary gear, results to a final drive of 15:1 to 10:1. It is used to compare results of ESVT use versus a single-stage transmission on the same vehicle [8]. A schematic diagram of the ESVT main components is shown in Figure 15.

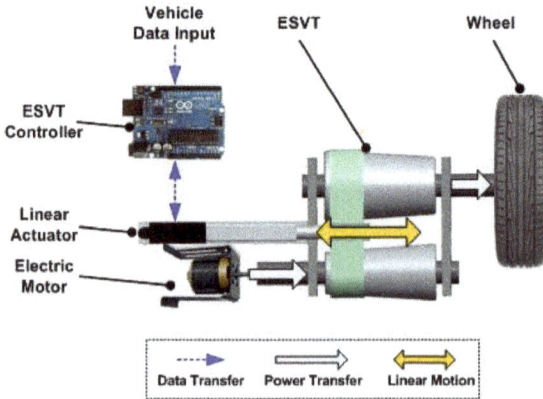

Figure 15. The proposed Electronic Shift Variable Transmission (ESVT) configuration.

Experimental results included a test case where the prototype vehicle operated in similar conditions (single lap test) at the same test area (Figure 8). Testing was performed initially without the ESVT and subsequently with it. The vehicle was accelerated at full throttle to obtain comparable results with limited driver intervention. The overall hydrogen consumption was lower using the ESVT (4.3% improvement). As already discussed, according to our literature review simulation results of CVT versus single stage transmissions showed 5–12.4% reduction on energy consumption for specific driving cycles. Since our experimental results show a 4.3% reduction in H_2 consumption for just one lap (corresponding to just one launch), it can be assumed that a very good correlation to simulation results exists.

3.4. Results Evaluation

There are several insights extruded by the experimental results presented. Regarding vehicle performance, the vehicle can achieve higher acceleration using the gearbox, corresponding to slightly lower time needed to complete the two laps in track. As shown in the comparison of Figures 9–14, the single stage transmission needs 83 s to complete the test, while the 2-speed gearbox times vary between 80–81 s. Time up to corner (1)–(2) is lower in every case of the 2-speed transmission. It was highlighted that hydrogen consumption can be reduced up to 3.4% (No gearbox: 4.43 lt H_2, 2-speed gearbox: 4.28 lt H_2), and mean power demand up to 3.8%. Even though this percentage seems low, it must be noted that tests conducted included just one launch during the laps. The same experimentation in a city driving cycle where often stop & go are necessary, may lead to further reduction of the energy consumption.

Another key finding was that gear change speed significantly impacts the energy consumption of an electric vehicle, and it can be reduced up to 4% or raised up to 11% (Table 2). As shown in Figure 16, there is lower energy demand on launch acceleration, but increased energy consumption in higher gear change speeds, since high electric motor rpm (in first gear) corresponds to motor poor efficiency region. In high vehicle speeds (with full throttle), increased power demand is found mainly due to gearbox losses. According to the above, a 2-speed gearbox can provide lower energy consumption but depends heavily on gear shift strategy. On the other hand, better vehicle performance is achieved.

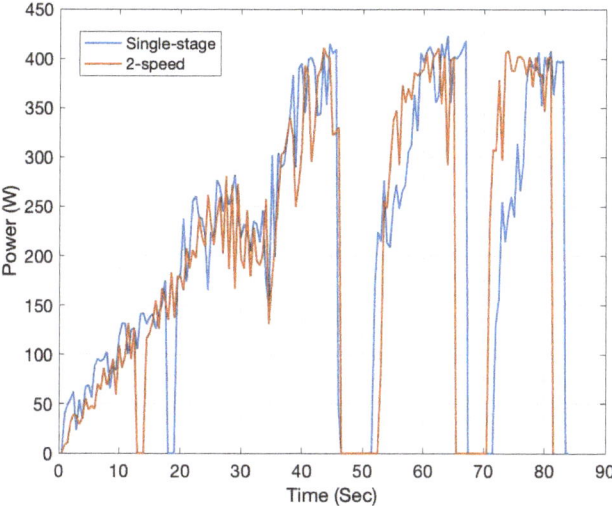

Figure 16. Comparison of power demand of the best gear change speed (11 Km/h) for a 2-speed gearbox setup versus a single-stage transmission.

Finally, the experimental results presented are compared to the simulated results found in literature [6,19,20]. It should be noted that to the best of our knowledge, there is no equivalent usage of on road testing with different transmission types and a detailed comparison of their effect. As so, according to literature, simulation results reveal that in a city driving circle, the energy consumed energy is reduced by 3.2–11% using a 2-speed versus single stage transmission. In our experimental results, 3.8% reduction is found in a 2-lap test. Thus, higher reduction should be expected for a city driving cycle that has more stop and go stages.

Regarding the CVT use, literature simulations indicated gains on energy consumption up to 12.5% for specific driving cycles. Our experimental results (using a prototype electronically controlled CVT) presented 4.3% energy reduction compared to single stage results, for just one lap test, so very good correlation to simulation results exists. It should be noted though that CVT drawbacks should be accounted for, such as weight, complexity and power losses.

All the above-mentioned conclusions point out that multiple stage transmissions use in electric vehicles can be an important factor to achieve lower energy consumption.

4. Conclusions

The on-road testing of different transmission using a custom electric testbed, powered by hydrogen fuel cell, is presented. The different configurations included a Standard Powertrain setup and Gearbox setup and a prototype 2-speed gearbox. We explored their performance as well as the effect of different gear change at different speeds. Experimental results showed that a 2-speed gearbox can provide lower energy consumption (up to 3.8%) but depends heavily on gear shift strategy. As such, gear change speed is of great importance and should be optimized to reach even higher levels of energy consumption reduction. Experimental results using a prototype electronically controlled CVT were also presented, indicating 4.3% energy reduction compared to single stage results, for just one lap test. All the experimental results presented for different transmissions use are compared to the simulated results found in literature. A good correlation is found regarding the use of single-stage, 2-speed and CVT transmissions.

Future work will be focused on new on road experiments that will reduce the driver's influence on results. This can be achieved by using the new platforms developed by the TUCER team, where throttle control can be defined either as a standard input of predefined values or using speed control over the track. Finally, simulation results will be provided based on the specific vehicle dynamic simulation, but validated through experimental testing, providing adequate results over different driving cycles.

Author Contributions: P.S. performed the original conceptualization, investigation, developed the methodology and contributed to the original manuscript writing. N.C.T. supervised the research effort and reviewed and edited the manuscript. L.D. formally analyzed the data, validated the experimental results and contributed to the original manuscript writing and editing. E.S.K. dealt with data curation and reviewed and edited the manuscript.

Funding: This work has been partially funded by the TUC's internal project "TUC Eco Racing team".

Conflicts of Interest: The authors declare no conflict of interest.

References

1. Chatelain, A.; Erriquez, M.; Mouliére, P.Y.; Schäfer, P. *What a Teardown of the Latest Electric Vehicles Reveals about the Future of Mass-Market EVs*; Technical Report; McKinsey: New York, NY, USA, 2018.
2. Miller, J. Hybrid electric vehicle propulsion system architectures of the e-CVT type. *IEEE Trans. Power Electron.* **2006**, *21*, 756–767. [CrossRef]
3. Ehsani, M.; Gao, Y.; Emadi, A. *Modern Electric, Hybrid Electric, and Fuel Cell Vehicles: Fundamentals, Theory, and Design*; Power Electronics and Applications; CRC Press: New York, NY, USA, 2009.
4. Kluger, M.A.; Long, D. *An Overview of Current Automatic, Manual and Continuously Variable Transmission Efficiencies and Their Projected Future Improvements*; SAE Technical Paper Series; SAE International: Warrendale, PA, USA, 1999; Volume 1259.
5. Spanoudakis, P. Design and Tuning of Operational Parameters for a Prototype Transmission System. Ph.D. Thesis, Technical University of Crete, Chania, Greece, 2013.
6. Holdstock, T. Investigation Into Multiple-Speed Transmissions for Electric Vehicles. Ph.D. Thesis, University of Surrey, Guildford, UK, 2014.
7. Bottiglione, F.; Pinto, S.D.; Mantriota, G.; Sorniotti, A. Energy Consumption of a Battery Electric Vehicle with Infinitely Variable Transmission. *Energies* **2014**, *7*, 8317–8337. [CrossRef]
8. Spanoudakis, P.; Tsourveloudis, N.C. On the efficiency of a prototype Continuous Variable Transmission system. In Proceedings of the 21st Mediterranean Conference on Control and Automation, Chania, Crete, Greece, 25–28 June 2013; pp. 290–295. [CrossRef]
9. Mantriota, G. Fuel consumption of a vehicle with power split CVT system. *Int. J. Veh. Des.* **2005**, *37*, 327–342. [CrossRef]
10. Zhang, Z.; Zuo, C.; Hao, W.; Zuo, Y.; Zhao, X.L.; Zhang, M. Three-speed transmission system for purely electric vehicles. *Int. J. Automot. Technol.* **2013**, *14*, 773–778. [CrossRef]
11. Qin, D.; Zhou, B.; Hu, M.; Hu, J.; Wang, X. Parameters design of powertrain system of electric vehicle with two-speed gearbox. *J. Chongqing Univ.* **2011**, *34*, 1–6.
12. Spanoudakis, P.; Tsourveloudis, N.C. A prototype variable transmission system for electric vehicles: Energy consumption issues. *Int. J. Automot. Technol.* **2015**, *16*, 525–537. [CrossRef]
13. Bingzhao, G.; Qiong, L.; Lulu, G.; Hong, C. Gear ratio optimization and shift control of 2-speed I-AMT in electric vehicle. *J. Mech. Syst. Signal Process.* **2015**, *50*, 615–631.
14. Sorniotti, A.; Pilone, G.L.; Viotto, F.; Bertolotto, S. A novel seamless 2-speed transmission system for electric vehicles: Principles and simulation results. *SAE Int. J. Engines* **2011**, *4*, 2671–2685. [CrossRef]
15. Spanoudakis, P.; Tsourveloudis, N.C.; Koumartzakis, G.; Krahtoudis, A.; Karpouzis, T.; Tsinaris, I. Evaluation of a 2-speed transmission on electric vehicle's energy consumption. In Proceedings of the 2014 IEEE International Electric Vehicle Conference (IEVC), Florence, Italy, 17–19 December 2014.
16. Elmarakbi, A.; Morris, A.; Ren, Q.; Elkady, M. Modelling and Analyzing Electric Vehicles with Geared Transmission Systems: Enhancement of Energy Consumption and Performance. *Int. J. Eng. Res. Technol.* **2013**, *2*, 1215–1254.
17. Sorniotti, A.; Subramanyan, S.; Turner, A.; Cavallino, C.; Viotto, F.; Bertolotto, S. Selection of the Optimal Gearbox Layout for an Electric Vehicle. *SAE Int. J. Engines* **2011**, *4*, 1267–1280. [CrossRef]

18. Rinderknecht, S. Electric power train configurations and their transmission systems. In Proceedings of the International Symposium on Power Electronics Electrical Drives Automation and Motion (SPEEDAM), Pisa, Italy, 14–16 June 2010; pp. 1564–1568.
19. Knödel, U.; Stube, A.; Blessing, U.C.; Klosterman, S. Design and Implementation of requirement-driven electric drives. *ATZ Worldw.* **2010**, *112*, 56–60. [CrossRef]
20. Ren, Q.; Crolla, D.; Morris, A. Effect of transmission design on Electric Vehicle (EV) performance. In Proceedings of the 2009 IEEE Vehicle Power and Propulsion Conference, Dearborn, MI, USA, 7–11 September 2009; pp. 1260–1265. [CrossRef]
21. Loveday, E. Vocis Says Two-Speed Electric Vehicle Transmission Will Boost Battery Life. Available online: https://www.autoblog.com/2011/08/19/vocis-says-two-speed-electric-vehicle-transmission-will-boost-ba/?guccounter=1 (accessed on 1 December 2018).
22. McKeegan, N. Antonov's 3-Speed Transmission for Electric Vehicles Boosts Efficiency by 15 Percent. Available online: http://www.gizmag.com/antonov-3-speed-transmission-ev/19088/ (accessed on 1 December 2018).
23. Kreisel Unveils an Automated 2-Speed Transmission for Electric Cars. Available online: https://electrek.co/2018/09/06/kreisel-automated-2-speed-transmission-electric-cars/ (accessed on 1 December 2018).
24. *Nexa^{TM} Power Module User's Manual*; Ballard Power Systems Inc.: Burnaby, BC, Canada, 2003.

© 2019 by the authors. Licensee MDPI, Basel, Switzerland. This article is an open access article distributed under the terms and conditions of the Creative Commons Attribution (CC BY) license (http://creativecommons.org/licenses/by/4.0/).

Article

Modeling a High Concentrator Photovoltaic Module Using Fuzzy Rule-Based Systems

Manuel Angel Gadeo-Martos [1,*], Antonio Jesús Yuste-Delgado [1], Florencia Almonacid Cruz [2], Jose-Angel Fernandez-Prieto [1] and Joaquin Canada-Bago [1]

1. Telematic Engineering System Research Group, CEATIC Center of Advanced Studies in Information and Communication Technologies, University of Jaén, Campus Científico-Tecnológico de Linares, C.P. 23700 Linares, Spain; ajyuste@ujaen.es (A.J.Y.-D.); jan@ujaen.es (J.-A.F.-P.); jcbago@ujaen.es (J.C.-B.)
2. IDEA Solar Energy Research Group, Center of Advanced Studies in Energy and Environment, University of Jaén, Campus las Lagunillas, C.P. 23071 Jaén, Spain; facruz@ujaen.es
* Correspondence: gadeo@ujaen.es

Received: 28 November 2018; Accepted: 31 January 2019; Published: 12 February 2019

Abstract: Currently, there is growing interest in the modeling of high concentrator photovoltaic modules, due to the importance of achieving an accurate model, to improve the knowledge and understanding of this technology and to promote its expansion. In recent years, some techniques of artificial intelligence, such as the Artificial Neural Network, have been used with the goal of obtaining an electrical model of these modules. However, little attention has been paid to applying Fuzzy Rule-Based Systems for this purpose. This work presents two new models of high concentrator photovoltaics that use two types of Fuzzy Systems: the Takagi-Sugeno-Kang, characterized by the achievement of high accuracy in the model, and the Mamdani, characterized by high accuracy and the ease of interpreting the linguistic rules that control the behavior of the fuzzy system. To obtain a good knowledge base, two learning methods have been proposed: the "Adaptive neuro-fuzzy inference system" and the "Ad Hoc data-driven generation". These combinations of fuzzy systems and learning methods have allowed us to obtain two models of high concentrator photovoltaic modules, which include two improvements over previous models: an increase in the model accuracy and the possibility of deducing the relationship between the main meteorological parameters and the maximum power output of a module.

Keywords: artificial neural network; fuzzy rule-based systems; adaptive neuro-fuzzy inference system; ad hoc data-driven generation; high concentrator photovoltaic modules; maximum power prediction

1. Introduction

Over the past few years, high concentrator photovoltaics (HCPV) have attracted the attention of multiple researchers. HCPV technology is based on the use of an inexpensive optical device to concentrate light in a little solar cell, typically a III-V multi-junction solar cell [1] with the aim of reducing the amount of expensive semiconductor material.

An HCPV module is the smallest unit able to transform non-concentrated solar radiation into electricity. It comprises solar cells, optical devices and peripheral components necessary to generate electricity. In addition, the HPCV module uses passive cooling to dissipate the great amount of heat that is generated when the solar cells are working at high concentration levels, and it also incorporates other components such as bypass diodes to prevent the overheating of cells.

As with any type of energy system, estimating the electrical characteristics of an HCPV module is going to be essential for designing, monitoring, life-cycle assessment and therefore promoting the market expansion of this emerging technology. Moreover, as with the conventional PV technology, the main parameters that influence the output of the HCPV module are solar spectrum, solar irradiance and temperature [1–7]. However, the modeling of an HCPV device, due to its special features, is more complex than the modeling of PV technology [8,9]. This is mainly due to the following factors:

(a) The use of high-efficiency multi-junction (MJ) solar cells, made of several p-n junctions of semiconductor material with different band gap energies, that are more influenced by changes in the solar spectrum than single-junction solar cells [10].
(b) The use of optical elements modifies the impact of the incident irradiance on the solar cells and introduces a strong angular dependence to the system [11,12].
(c) The difficulty of measuring the temperature at which the solar cells are working in an HCPV module; since they are mounted on a substrate surrounded by other peripheral elements, the direct measurement of this temperature is not possible without deleterious the module or the components of the assembly that surround the cells [13–15].
(d) The HCPV modules only react to the direct component of the solar radiation (DNI) due to the use of point-focus optical elements. This component of the solar radiation is more variable and difficult to predict than the global irradiance since it is more affected by the existence of clouds and aerosols in the atmosphere [10,16].

Consequently, the modeling of HCPV systems is complex and still challenging from a fundamental point of view in many cases. As of today, several models of HCPV modules have been developed [9,17–30]. Most of these models try to determine the relations that exist among the electrical parameters of an HCPV module (i.e., open-circuit voltage, short-circuit current and maximum power) and the main atmospheric parameters that have influence on them. One of the researched approaches to address this issue is based on the use of artificial intelligence techniques, mainly artificial neural networks (ANN) [31]. This is due to the advantages offered by the ANNs to solve non-linear and complex problems and the great level of complexity of electrical modeling of HCPV devices, as discussed above. The use of ANNs has the benefit of offering alternative solutions to complex problems that are still challenging from a fundamental physical approach due to the complexity of the different physical event involved in the performance of these systems.

Taking into account that, in recent years, the Fuzzy Rule-Based Systems (FRBS) [32] have been successfully used to model a large number of systems [33–35], in this paper we propose the use of FRBS with the goal of modeling HCPV systems. For that purpose, two types of FRBS have been proposed: a) the Takagi-Sugeno-Kang (TSK) systems [36], which are characterized by the achievement of high accuracy in the model, and b) the Mamdani systems [37,38], which is characterized by high accuracy and the ease of interpreting the linguistic rules that control the behavior of the fuzzy system.

To generate a good model based on an FRBS, it is necessary to have a good knowledge base. To obtain such a model, in this paper we propose two methods: a) "Adaptive neuro-fuzzy inference system" (ANFIS) [39] and b) "Ad Hoc data-driven generation method" [40–42]. These methods have been used in the process of linguistic rule learning, in TSK and Mamdani FRBS, respectively. Finally, the experimental results show that the new HCPV models proposed in this paper improve the accuracy of the best published model [8].

The remainder of the article is organized as follows: In Section 2, we introduce the related work, and in Section 2.1, a review of the models for the electrical characterization of HCPV is presented. In Section 2.2, a short description of the architecture and reasoning method of the ANN is given. In Section 2.3, a summary of the applications of ANNs in the field of HPCV is presented. Section 2.4 describes an HCPV model based on ANN. Section 3 demonstrates the utility of the FRBS for modeling an HCPV module. In Section 3.1, we give a description of the FRBS concept, the main types and their characteristics with respect to the structure of the fuzzy rules, as well as an introduction to the design process. Sections 3.1.1 and 3.1.2 describe the types of fuzzy systems proposed in this paper to model the HCPV modules: Mamdani FRBS and TSK FRBS. In Section 3.2, the learning methods used to define the fuzzy rules that control the behavior of the FRBS are introduced. Section 3.2.1

presents the ANFIS method. Section 3.2.2 shows the Ad Hoc data-driven generation method. Section 4 describes the new methods proposed in this paper to improve the model of the HCPV module, and in Section 4.1, we discuss the particularization of the methodology presented in Section 3.2.1 to obtain a good HCPV model based on a TSK FRBS, and in Section 4.2, we discuss the particularization of the methodology proposed in Section 3.2.2 to obtain a good HCPV model based on a Mamdani FRBS. In Section 5, we show the results obtained by applying the methodologies presented in Sections 4.1 and 4.2. In Section 6, we extend the results of Section 5.

2. Related Work

2.1. Models for the Electrical Characterization of HCPV

In [43] a critical review of the models for the electrical characterization of HCPV has been made. In this paper, it shows that the model accuracy it is related with the number of "physical sensor", thus the ATSM [18] and the King [20] models, that only take in count three atmospheric parameters, presents a medium accuracy level. On the other hand there are others models as Chan [28] and Steiner [30] characterized by an advanced level of accuracy but which use a complex physical model and whose parametrization require advanced module information and indoor measurements as well as require specific software.

In [19] Peharz et al. proposed a model that provides better accuracy than the previous ones but it requires a more expensive experimental set-up. In [9] Fernandez et al. proposed a simple model, with an optimum balance between simplicity and accuracy, based on simple mathematical relationships between atmospheric parameters, but this model is focused for sites with climates characterized by high-medium turbidity levels.

In [29] Almonacid et al. proposed a model that use of the ability of artificial neural networks in order to obtain the complex relations between the electrical behavior of HCPV module and their inputs variables, with a high level of accuracy. However, its difficulty to be applied is that a deep knowledge on ANN architectures and its training algorithms is necessary to determine the parameters of this model, and this is not always available to be applied.

Energy prediction is a key factor in the market integration and growth of any type of energy production system. However, due to their special features, the modeling of HCPV devices is quite complex and still challenging. As commented, one of the more important lines of research to try to address some of the issues related to this technology is based on the use of artificial intelligence techniques, mainly ANNs. Therefore the next subsections present a brief summary of ANN and its application in the field of HCPV.

2.2. Artificial Neural Networks

Artificial neural networks (ANN) are a widely used tool for approximating nonlinear functions that have multiple entries [44]. An artificial neural network simulates the behavior of the dendrites and axons of nerve cells. In particular, the system inputs are related to the outputs through different neurons grouped in layers that are joined together by arcs. These arcs are denoted mathematically by a number or weight. The type of neural network that is often used in the literature is called the multilayer perceptron. Neurons are grouped into three types of layers: input, output and intermediate layers (called hidden layers). The number of neurons in each layer and the number of hidden layers determine the complexity of the artificial neural network.

2.3. HCPV and ANNs

ANNs have been used for modeling some issues related to MJ solar cells; for instance in [45–47] several ANN-based models have been proposed for estimating the tunneling effects, the the I-V characteristic and External Quantum Efficiency (EQE), both under one sun and in darkness, for dual-junction (DJ) GaInP/GaAs solar cells. In [48] the authors used an ANN-based model to estimate the performance of triple-junction (TJ) InGaP/GaAs/Ge solar cells and the EQE under the influence of an ample range of charged particles. Finally, in [49] Fernández et al. presented three ANN-based models of the main electrical parameters of a TJ solar cell: open-circuit voltage (V_{oc}), short-circuit current (I_{sc}) and maximum power (P_{max}).

Regarding HCPV modules and systems, several ANN-based models have been developed to try to solve some problems related to these devices. In [50] Fernandez et al. presented an ANN-based model for spectrally corrected direct-normal irradiance. This parameter is defined as the part of the incident spectrum that an HCPV module is capable to convert into electricity, so it can be used to quantify the spectral effects on an HCPV device. In [13] and [51] two ANN-based models to estimate the cell temperature of an HCPV module were developed. The direct measurement of this temperature is quite complex since it is not possible to access the inside of the module without damaging it. ANN-based models were developed to try to solve this problem by attempting to characterize the relationship between the main meteorological parameters and the cell temperature that affect the temperature. In [29] and [52] two ANN-based models were developed to estimate the maximum power of an HCPV module: a feed-forward artificial neural network and a cooperative-competitive hybrid algorithm for radial basis-function networks. Finally, in [53,54] two ANNs were used to model the complete I-V characteristic of an HCPV module.

2.4. Application of ANNs to Obtain an HCPV Model

The solar research group of Jaén University has ample experience in the application of ANN in the photovoltaic field [55–59]. Taking into account this knowledge, Almonacid et al. [29] developed an ANN-based model to find the relation between the main parameters that affect its performance: spectrum (S), irradiance (B), wind speed (Ws) and temperature (T) and the output of an HCPV.

The architecture of the proposed ANN was based on a three-layer feed-forward neural network trained with the Levenberg-Marquardt back-propagation algorithm. The input layer had five nodes: the direct-normal irradiance (DNI), the air mass (AM), the precipitable water (PW), the wind speed (Ws) and the air temperature (Tair). The ANN-based model used the precipitable water and air mass to evaluate the spectrum, the direct-normal irradiance to evaluate the irradiance and the wind speed and the air temperature to evaluate the cell temperature. The hidden layer had seven nodes, and there was one node in the output layer (P_{max}). Coefficients of the neural network were obtained from outdoor monitoring. The results show that the ANN-based model could be used to calculate successfully the output of an HCPV module with an RMSE of 2.91%, an MBE of 0.07% and an R^2 of 0.99.

By virtue of this results it may be stated that this model provides one of the best level of accuracy between the models mentioned in Section 2.1. However, as commented, the difficulty in applying this model is focused in the determination of the parameters, which requires a deep knowledge of complex artificial neural networks architectures and their training algorithms. On other hand, once the parameters have been determined, the knowledge remains implicit and the model becomes easy to use, however the complex relations that link the HCPV module behavior and the atmospheric variables have a short degree of interpretability. In order to improve the knowledge over these relations without decrease the level of accuracy in this paper we propose the use of FRBS.

3. Utility of the FRBS for Modeling an HCPV Module

To obtain the maximum annual electric power for an HCPV system, it is advisable to use highly accurate and stable solar trackers. High-accuracy dual-axis solar trackers, which track the Sun's motion across the sky, must be used in HCPV systems because these systems only accept direct solar light with a very small deviation in the acceptance angle [60].

In the literature, we can observe the use of FRBS to implement solar tracker using fuzzy control [61–64]. In [65], the authors explain that fuzzy control could be a good option when an accurate model of the tracking system is absent. Therefore, if we use this type of control for a solar tracker, it is necessary to learn a good knowledge base (KB) for this control system, and to achieve this task it is necessary to have a good model of the HCPV module.

Because of the large quantity of relevant variables, the problem of modeling a HCPV becomes very complex, thus it is advisable to solve it by means of modeling techniques able of obtaining a model representing the non-linear relationships existing in it. Furthermore, the problem-solving goal must be to obtain a user-interpretable model, as well as obtain an accurate model, able of putting some light on the relations that exist among the electrical parameters of an HCPV module and the main atmospheric parameters that have influence on them. Due to all these reasons, in this paper we introduce the use of FRBSs in order to model the HCPV module.

A Fuzzy Rule-Based System is any system, where Fuzzy Logic may be used either to model the relationships and interactions among the system variables, or as the basis for the representation of different forms of knowledge. Fuzzy Systems have proven to be a significant tool for modeling complex systems in which, due to the imprecision or the complexity, conventional tools are unsuccessful [33,34,66,67]. For that reason, they have been successfully used to a wide range of problems from different areas presenting vagueness and uncertainty in different ways [68].

With the general purpose of optimizing the HCPV fuzzy control system, in this paper we introduce two new HCPV module models. Each is a variation of a Fuzzy Rule-Based System and is characterized by their properties. For this reason, the sections below detail the study of Fuzzy Rule-Based Systems, their types and properties.

3.1. Fuzzy Rule-Based System

In 1965, Zadeh [69] suggested a modified set theory, called fuzzy set theory, in which an element could have a degree of membership that takes continuous values, rather than being 0 or 1. One of the most important areas of application of his theory is the FRBSs. An FRBS is a rule-based system in which fuzzy logic (FL) is used as a tool for managing different forms of knowledge about the problem at hand [32]. These systems are a development of the classical rule-based system because they address "*IF-THEN*" rules whose consequents and antecedents are composed of fuzzy statements (fuzzy rules), instead of classical logical statements. FRBSs incorporate the human knowledge of the expert using the FL. One of the main characteristics of these systems is the ability to incorporate human knowledge by means of uncertainty or imprecision and lack of accuracy.

FRBSs have some advantages over traditional rule-based systems, such as (a) inference methods become more robust with the approximate reasoning methods employed within FL and (b) the key features of knowledge captured by fuzzy sets involve the handling of uncertainty.

In an FRBS, the knowledge is stored in a knowledge base that consists of three elements: membership functions, a set of "*IF-THEN*" rules, and linguistic variables. These rules are defined by their consequents and antecedents. Antecedents and, frequently, consequents, are associated with fuzzy concepts.

Because of these properties, FRBSs have been successfully applied to modeling the interactions and relationships that exist between their input and output variables. In this way, one of the most common application of FRBSs is fuzzy modeling [34]. There are two types of FRBSs for engineering problems:

(a) *Mamdani FRBSs* [37]. In this type of FRBSs, antecedents and consequents are composed of linguistic variables. The rules possess the following form.

IF X_1 is A_1 and ... and X_n is A_n THEN Y is B,

where X_i are input variables, A_i are fuzzy sets associated to input variables, Y is the output variable and B is a fuzzy set associated to the output variable.

(b) *Takagi-Sugeno-Kang FRBSs* [36]. This model of FRBSs is based on rules in which the antecedent is composed of linguistic variables and the consequent is an analytical function of the input variables. In this case, the rules are as follows:

IF X_1 is A_1 and ... and X_n is A_n THEN $Y = f(X_1, \ldots, X_n)$.

In most cases, the function is a linear function:

IF X_1 is A_1 and ... and X_n is A_n THEN $Y = p_0 + p_1* X_1 + \ldots + p_n* X_n$, with X_i as the input variables, Y as the output variable, A_i as the fuzzy sets related to the input variables, and $p = (p_0, p_1, \ldots, p_n)$ as a vector of real parameters. These types of rules are usually called TSK fuzzy rules.

The accuracy of an FRBS directly depends on two aspects: the composition of the fuzzy rule set and the way in which it implements the fuzzy inference process. Therefore, the FRBS design process includes two main tasks [32]:

(a) The choice of the type of FRBS (Mamdani or Takagi-Sugeno-Kant), and the different fuzzy operators that are employed by the inference process.
(b) The generation of the fuzzy rule set. That requires some design tasks, such as

1. Selection of the relevant input and output variables.
2. Definition of the scale factor, the number of term sets and the membership function for each linguistic variable.
3. Derivation of the linguistic rules that will form part of the rule set.

With the intention of use the best properties and facilities provided by the FRBSs, in this paper we propose the generation of two HCPV models based on a Mamdani FRBS and a Takagi-Sugeno-Kant (TSK) FRBS. This systems are explained in more detail below, as well as the adaptations made in our proposal.

3.1.1. Mamdani Fuzzy Rule-Based Systems

In 1974, Mamdani [37] proposed an FRBS that addresses real inputs and outputs. He augmented Zadeh's formulation to apply an FS to a control problem. Later, in 1975, he proposed the Fuzzy Logic Controllers (FLCs) [38] as a type of FS, which are referred to as FRBSs with fuzzifier and defuzzifier. Hence, FLCs may be considered knowledge-based systems, that incorporate human knowledge into their Knowledge Base through fuzzy membership functions and fuzzy rules. As can be observed (see Figure 1), a Mamdani FRBS consists of the following components [32]:

(a) A Knowledge Base (KB), which stores the knowledge about the problem. The KB comprises a Database (DB) and a fuzzy Rule Base (RB). The DB contains the definitions of the fuzzy rules and linguistic labels, and the RB comprises the collection of linguistic rules representing the expert knowledge.
(b) A fuzzy inference engine, that presents the following structure:

1. A fuzzification interface, which converts the values of input variables into fuzzy information, for this it assigns grades of membership to each fuzzy set defined for that variable.
2. An inference system, which infers fuzzy outputs by employing fuzzy implications and the rules of inference of FL.
3. A defuzzification interface, which produces a non-fuzzy output from an inferred fuzzy output. This interface has to aggregate the information provided by the output fuzzy sets and to obtain an output value from them. There are two basic techniques for doing this: Mode B-FITA (first infer, then aggregate) and Mode A-FATI (first aggregate, then infer). The FRBS proposed in this paper uses mode B-FITA because it reduces the computational burden compared to mode A-FATI.

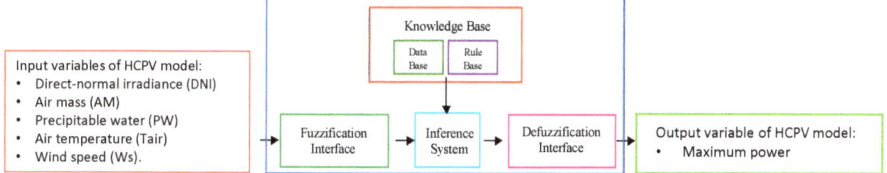

Figure 1. Structure of a HCPV model based on a Mamdani FRBS.

In the same way as the ANN-based model, the input variables of the proposed HCPV models based on a FRBS are: the direct-normal irradiance (DNI), the air mass (AM), the precipitable water (PW), the air temperature (Tair) and the wind speed (Ws), as well as the output variable is the maximum power.

In this model based of a Mamdani FRBS we have chosen a uniform partition of the membership functions into the fuzzy sets of output and input variables. Figure 1 shows the structure of a HCPV model based on a Mamdani FRBS.

3.1.2. Takagi-Sugeno-Kang Fuzzy Rule-Based Systems

The output of a TSK FRBS is the sum of the individual outputs provided by each rule, Y_i ($i = 1, \ldots, m$, where m is the number of rules of the TSK KB), weighted by the values of the parameters h_i, where $h_i = T(A_{i1}(x_1), \ldots, (A_{in}(x_n)))$ is the matching degree between the current inputs to the system, $x_0 = (x_1, \ldots x_n)$ and the antecedent part of the i-th rule. T stands for a conjunctive operator modeled by a t-norm. To design the inference engine of a TSK FRBS, the designer only must select this connective operator T. The most common choices are the algebraic product and the minimum [36] (In this paper the algebraic product have been used as an operator T). Figure 2 shows the structure of a HCPV model based on a TSK FRBS.

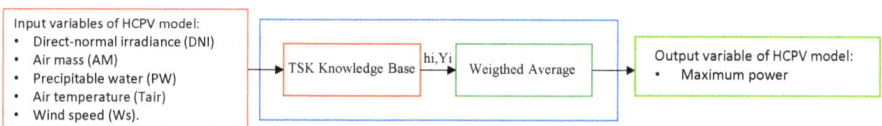

Figure 2. Structure of a HCPV model based on a TSK FRBS.

To help achieve the goals of both Mamdani FRBSs and TSK FRBSs they must have a complete and consistent knowledge base, therefore it is necessary to use learning methods to define the fuzzy rules.

These models are capable of estimate the HCPV maximum power from outdoor measurements over atmospheric variables, but they are not suitable to model the I-V curve and no to evaluate algorithms to obtain maximum power point tracking (MPPT), because they are complex processes, that require modelling the HCPV module with a number of curves sufficiently representative, in which each one of their parameters will be influenced by each one of the atmospheric parameters.

3.2. Learning Methods to Define the Fuzzy Rule Set

To define the fuzzy rule set in modeling applications of FRBSs, two types of information are available to the designer: linguistic and numerical. Related to this information, there are two main ways of deriving the fuzzy sets:

Derivation from the experts. In this case, the human expert specifies the linguistic labels, the structure of the rules and the meaning of each label. This method is useful only if the expert is capable to express his knowledge in the form of linguistic rules.

Derivation from automated learning methods based on the existing numerical information. To overcome the difficulties related with the derivation of the knowledge from the expert, in recent years, numerous inductive methods have been developed. Such methods include ad hoc data-driven generation methods [40–42], neural networks [44], clustering techniques [70,71], and evolutionary algorithms [72,73].

For modeling a system with very high complexity (such as HCPV), it is very difficult to obtain the linguistic rules that define the behavior, and therefore in this paper we propose the use of two learning methods to define the fuzzy rule set: (a) ANFIS and (b) the Ad Hoc data-driven method.

3.2.1. ANFIS

ANFIS is capable of providing in a single tool the advantages of systems based on neural networks and fuzzy systems. Thus, the expert knowledge provided by the fuzzy rules is complemented by the training and validation of neural networks. Researchers have frequently used ANFIS since it was first presented by Jang et al. in [39] for different problems such as pattern classification or adjustment of non-linear functions. This method has been used for discriminating voice/music [74], for modeling solar radiation [75], to predict air pollution [76], in solar PV [77], and in treating water qlity [78], to cite just a few current examples.

ANFIS uses a set of data (input/output) to build a TSK fuzzy system [39] in which the parameters of member functions and the relations with the output are adjusted using neural networks. For this purpose, a fuzzy system is identified with a neural network, as shown in Figures 3 and 4.

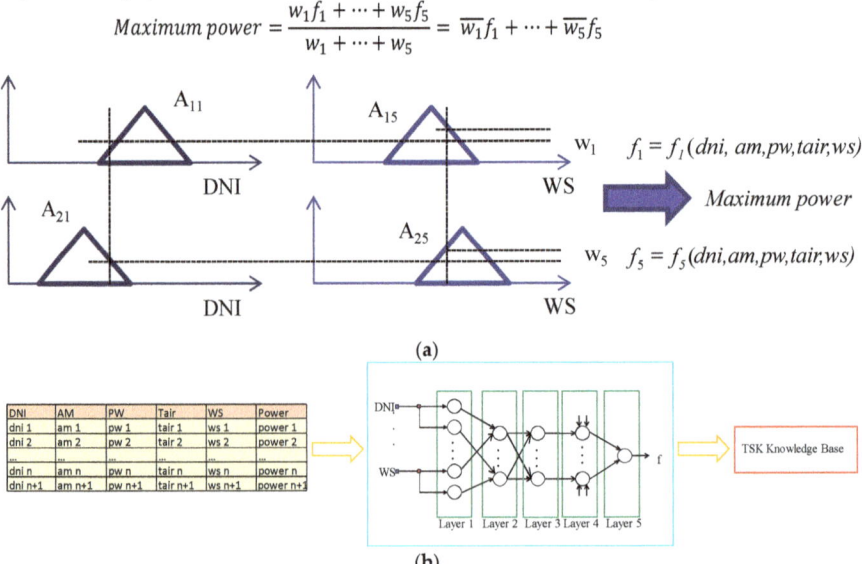

Figure 3. ANFIS versus ANN: relationship. (**a**) Antecedent and consequent part of fuzzy rule; (**b**) Inputs, ANN and KB of ANFIS structure.

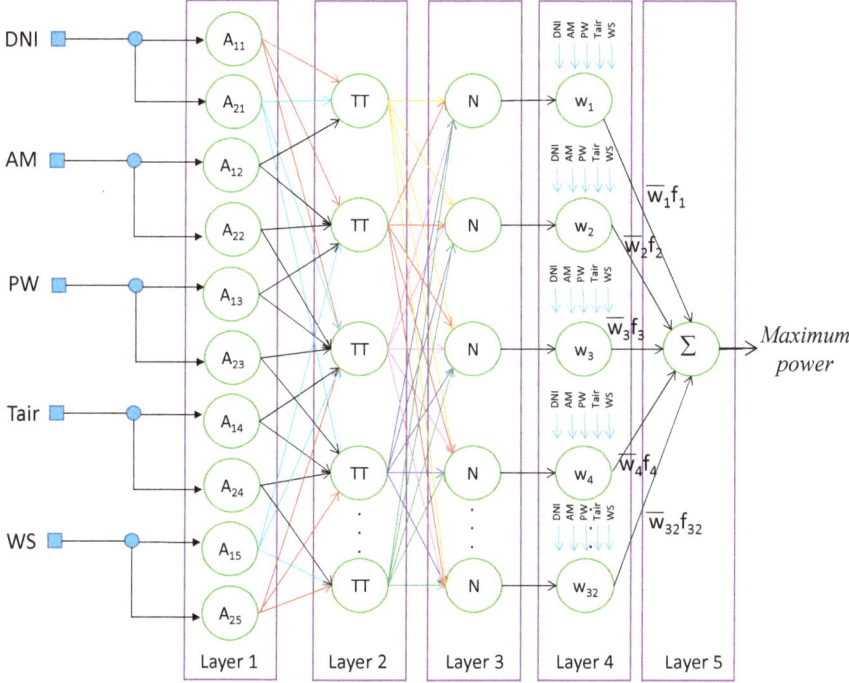

Figure 4. Detail of ANN used in the ANFIS method.

In Figure 3, at the top, the inputs are fuzzified into the system by means of membership functions. Then, the output from the combination of the IF-THEN rules is obtained. A five-layer neural network, as illustrated in Figure 4, can model this process. Each layer of the ANFIS model handles a task:

(a) Layer 1: Responsible for determining the degree of membership of each input to each of the fuzzy sets (fuzzification of inputs).
(b) Layer 2: The output of this layer is the products obtained by the rules that are activated. Each node symbolizes the firing strength of the rule (wi).
(c) Layer 3: Outputs of this layer are denominated normalized firing strengths (wi). The ith node obtains the ratio of the ith rule's firing strength to the addition of all rule's firing strengths.
(d) Layer 4: Outputs are obtained from the consequent parameters of the rules (fi).
(e) Layer 5: Calculates the overall output as the sum of all incoming data.

Through a process of training, you can obtain all the values of the different layers and then create the fuzzy knowledge base.

3.2.2. Ad Hoc Data-Driven Generation Method

A family of simple and efficient methods, called ad hoc data-driven methods, which generate new rules from the training database, guided by the covering criteria of the data in the example set, have been proposed in the literature [40–42]. The ad hoc data-driven linguistic rule-learning methods are identified by four main features:

(a) They are based on working with an input-output data set.
(b) They consider a previous definition of the database, composed of the output and input primary fuzzy partitions.

(c) The generation of the linguistic rules is guided by the covering criteria of the data in the example set.
(d) The learning mechanism is not based on search techniques or any well-known optimization; it is specifically developed for this purpose.

In this paper, we propose the use of the simpler type, the "ad hoc data-driven linguistic rule-learning methods guided by examples", to obtain the linguistic rules in the Mamdani FRBS. This method obtains each rule from a specific example in the data set. These rules belong to a candidate rule set because after this generation stage, a selection process is performed to derive the final RB.

One of the most widely used and well-known example-based methods is Wang and Mendel's method [41]. This method implements the RB generation by means of the following steps (see Figure 5):

(a) Consider a fuzzy partition of the variable space.
(b) Generate a candidate linguistic rule set. This set will be composed by the rules best covering each example included in the input-output data set.
(c) Assign an importance degree to each rule. This will be obtained by computing the covering value of the rule.
(d) Obtain a final RB from the candidate linguistic rule set, by selecting the rule with the highest importance degree in each group.

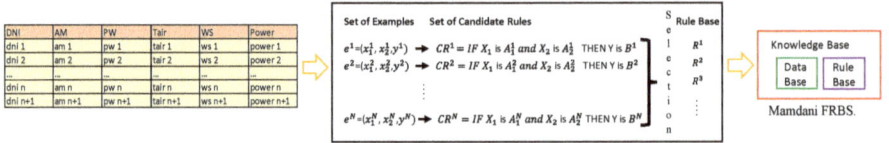

Figure 5. Graphical illustration of the rule-generation process.

4. Proposed Methods to Obtain an HCPV Model

4.1. Application of ANFIS to Train a TSK FRBS to Obtain an HCPV Model

The proposed ANFIS model has the following characteristics:

(a) The five variables mentioned above as inputs: direct-normal irradiance (DNI), wind speed (Ws), air temperature (Tair), precipitable water (PW), and air mass (AM), each with two Gaussian fuzzy sets.
(b) The output variables as linear functions: maximum power.
(c) A training algorithm based on back-propagation.

4.2. Application of Ad Hoc Data-Driven Methodology to Obtain an HCPV Model

In this case, we use a Mamdani FRBS to provide a highly flexible method to formulate knowledge, while at the same time remaining interpretable. In these systems, each rule is a description of a relation between one condition and one action, which allows for easy interpretation by a human. This property makes these systems useful for application in problems that require high model interpretability, such as fuzzy control [79], and linguistic modeling [34,35].

To obtain a model of the HCPV module behavior, we select the following input variables: DNI, AM, PW, Tair, and Ws. These variables have been normalized using a linear transformation that was applied between their minimum and maximum values. The output will be the estimated maximum power delivered by the HCPV module. Taking into account the steps described in Section 3.2.2, we implement the following steps:

(a) A fuzzy partition of the input and output variable space is created. Over all of the five input variables, we have applied a uniform partition composed of 19 fuzzy sets, and over the output variable we have applied a uniform partition with 495 fuzzy sets. Based on these values, we have generated a partition of the five-dimensional space that is composed of 19^5 subspaces or regions.
(b) Candidate rules are generated in each fuzzy input subspace. For each example of the training set (which is composed of 8926 examples) contained in each subspace, a fuzzy rule is proposed. All of these rules have the same antecedent and differ in their consequents.
(c) Each rule is assigned an importance degree, which is obtained by computing the covering value of the rule in its subspace [42].
(d) Select the rule with the highest importance degree in each group or subspace. The algorithm generated a KB with 6216 different rules.

In this learning algorithm, the accuracy of the modeled output variable is related to the number of subspaces into which the original five-dimensional space was divided. In the same way, the complexity of the model increases with the number of subspaces and their dimension. On the other hand, the number of rules, which is related to the number of subspaces, determines the accuracy of this model.

5. Experimental Results

As commented, the ANN model provides one of the best level of accuracy between the models mentioned in the related work. For that reason, in this paper, the developed models are evaluated and compared to the ANN model.

As a result of executing these proposed modeling methods (see Sections 4.1 and 4.2), we have obtained two models of an HCPV:

(a) TSK FRBS training with ANFIS.
(b) Mamdani FRBS obtained with the Ad Hoc data-driven method.

To conduct this study, the electrical performance of a HCPV module mounted on a two-axis solar tracker (Figure 6 left) together with the main atmospheric parameters that have influence in its output, have been measured at the Centro de Estudios Avanzados en Energía y Medio Ambiente (CEAEMA) of the University of Jaén.

Figure 6. Experimental set-up used to conduct this study at the Centro de Estudios Avanzados en Energía y Medio Ambiente at the University of Jaén. (**left**) HCPV module; (**right**) atmospheric station.

The module is made up of 20 triple-junctions lattice-matched GaInP/GaInAs/Ge solar cells interconnected in series. The module uses silicon-on-glass (SOG) Fresnel lenses as primary optical element. The secondary optical element consists of reflexive truncated pyramids made up of an aluminium film layer to enhance the reflectivity. The module has an optical efficiency of 80%, a geometric concentration of 700 and uses passive cooling to ensure that MJ solar cells operate on their optimal operation range [49]. Table 1 shows the main electrical characteristics of the HCPV module under Concentrator Standard Operating Conditions, CSOC, (i.e., AM1.5D, DNI = 900 W/m², ambient temperature, T = 20 °C and wind speed, WS = 2 m/s).

Table 1. Electrical characteristics of the HCPV module under Concentrator Standard Operating Conditions, CSOC, (i.e., AM1.5D, DNI = 900 W/m², ambient temperature, T = 20 °C and wind speed, WS = 2 m/s).

Electrical Characteristics	Value
Short-circuit Current (I_{sc})	5.30 A
Open-circuit Voltage (V_{oc})	57.28 V
Maximum Current (I_{ppm})	4.85 A
Maximum Voltage (V_{ppm})	47.62 V
Maximum Power (P_{max})	230.85 W

In order to measure the electrical parameters of the HCPV module a four-wire electronic load located was used. In addition, the centre is equipped with an atmospheric station MTD 3000 from Geonica S.A. (Figure 6 right) to record the main atmospheric parameters such as direct normal irradiance, air temperature, wind speed or humidity. The values of aerosol optical depth not provided by the atmospheric station were obtained from MODIS Daily Level-3 data source [80].

To obtain the knowledge base that characterizes each model, in this paper, we have considered a set of input and output variable values, recorded by the experimental setup, that it is composed of 11155 examples that were recorded every 5 minutes from January 2011 to December 2012.

In the case of ANFIS, to prevent the learning method from memorizing the training examples, the available data are separated into three subsets:

(a) A training set: the learning method is trained with this set in the classical way. It includes 33% of the examples.
(b) A validation set: each time you complete a training iteration, this set is used to determine whether you can stop training. The goal is to avoid overtraining. This set contains 33% of the examples.
(c) A test set to discern the goodness of the process. This set contains 33% of the examples.

After this, the following procedure was used. The system trains on the first set for 10 iterations. Then the validation set is used, and if a result with low enough error is obtained, the process is finished; otherwise another 10 iterations are performed. Finally, the third set is used to check the degree of fitness of the system. The three sets each contain 33% of the samples.

In the case of the Ad Hoc data-driven method, to prevent biasing the learning process, the available data are divided into two subsets:

(a) A training set, which includes 80% of the examples.
(b) A test set, which contains 20% of the examples.

This training set has been used to generate the linguistic rules that constitute the KB. Then the test set was employed to evaluate the performance of this learning method.

All the earlier sets have been obtained based on all data sets and were formed by carrying out a random and uniform selection of the subspaces, and then a random selection of an example contained in each subspace.

As a result of executing the proposed modeling methods, the characteristics of the obtained knowledge bases are as follows:

(a) For TSK, FRBS training with ANFIS:

   1. Number of rules: 32.

2. Number of fuzzy sets for each input variable: 2.
3. The partition of the membership functions into the fuzzy sets of input variables is shown in Figure 7. From the observation of the captured data and the membership functions of the two fuzzy sets, for each of the atmospheric variables represented in Figure 7, the conclusion can be drawn that the crossing point of the two fuzzy sets has been chosen so that the number of samples is approximately equal to right and left of said point. Therefore, the high and low fuzzy sets will cover approximately the same number of samples, half of the total number of samples captured. This distribution of the membership functions, which responds to the density of the samples captured, as well as the choice of the initial and final values of the height of the fuzzy sets and the slope of the curve that defines them, allows to have a number of very reduced rules, without a decrease in model accuracy.

(b) For Mamdani, FRBS obtained with the Ad Hoc data-driven method:

1. Number of rules: 6216.
2. Number of fuzzy sets for each input variable: 19.
3. Number of fuzzy sets for each output variable: 495.
4. Uniform partition of the membership functions into the fuzzy sets of output and input variables.

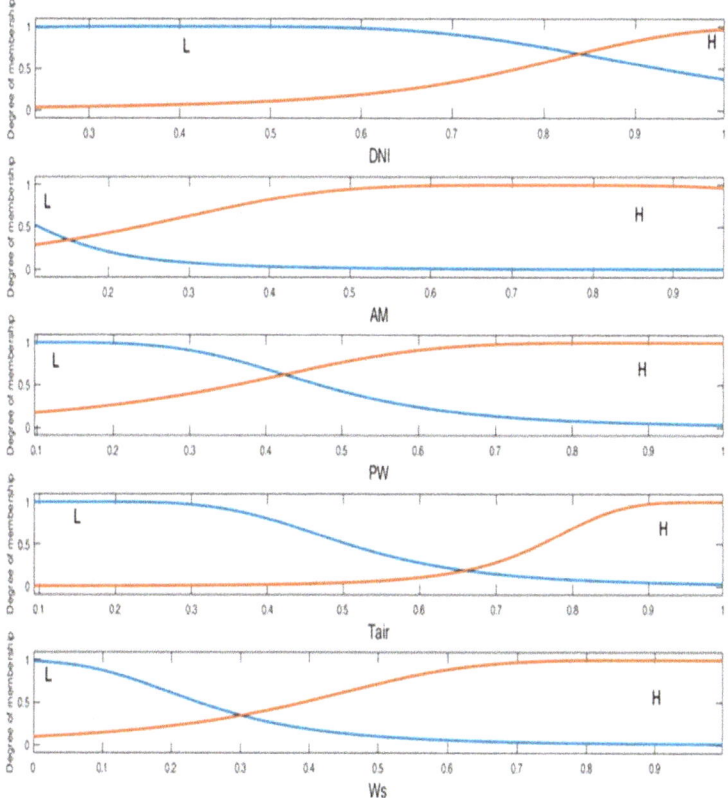

Figure 7. Membership functions of the input variables for the HCPV model based on the TSK FRBS training with the ANFIS method.

After the KBs have been obtained, in order to evaluate the level of accuracy of these models, for each element of the test set (five values of atmospheric variables) it is made the simulation of each model to obtain the predicted maximum output power. To study the behavior of these models, we have to calculate the normalized root mean square error (NRMSE) between the measured (Pm) and the predicted (Pp) maximum output power. The equations of the RMSE (in watts) and NRMSE (in %) are:

$$RMSE = \sqrt{\frac{\sum_{i=1}^{n}(P_p - P_m)^2}{n}} \quad NRMSE = \frac{RMSE}{\overline{P_m}} \quad (1)$$

The values of NRMSE (in %), obtained with the three models of HCPV (including the ANN model introduced in [29]) are shown in Table 2. A comparison of the values of the NRMSE gain for the models of HCPV presented in this paper versus the ANN model, is shown in Table 3. The equations of the gain in NRMSE are:

$$Gain_{NRMSE\ TSK\ FRBS\ vs\ ANN} = \frac{|NRMSE_{TSK\ FRBS} - NRMSE_{ANN}|}{NRMSE_{ANN}} \quad (2)$$

$$Gain_{NRMSE\ Mamdani\ FRBS\ vs\ ANN} = \frac{|NRMSE_{Mamdani\ FRBS} - NRMSE_{ANN}|}{NRMSE_{ANN}} \quad (3)$$

Table 2. NRMSE for the three models: ANN, TSK FRBS training with ANFIS and Mamdani FRBS obtained with Ad Hoc data-driven method.

NRMSE (%)	Train	Validation	Test	Overall
ANN	2.11	2.1	2.45	2.16
TSK FRBS training with ANFIS	1.85	1.99	2.21	1.93
Mamdani FRBS training with Ad Hoc data-driven	2.04		2.25	2.07

Table 3. Comparison of the gain in NRMSE, for the model TSK FRBS training with ANFIS versus ANN and for Mamdani FRBS obtained with the Ad Hoc data-driven method versus ANN.

Gain in NRMSE	Train	Validation	Test	Overall
ANN	0	0	0	0
TSK FRBS training with ANFIS	12.32	5.24	9.80	10.65
Mamdani FRBS training with Ad Hoc data-driven	3.32		8.16	4.17

In addition to the NRMSE analysis between the measured and predicted output maximum power, the linear regression analysis have also been calculated to study the results of the compared methods (ANN, Mamdani, TSK) in more detail, as shown in Figures 8–10.

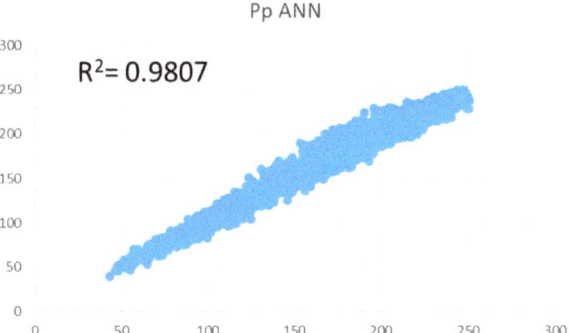

Figure 8. A regression analysis between the predicted output maximum power in the ANN model and the measured output maximum power for the total set of data.

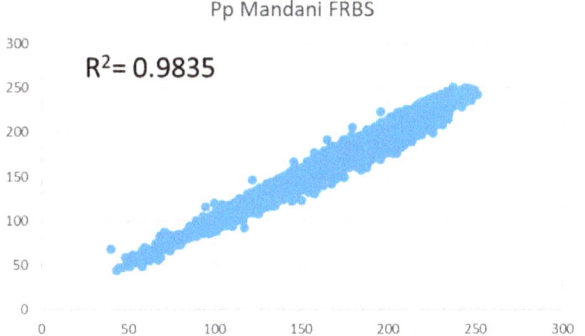

Figure 9. A regression analysis between the predicted output maximum power in the Mamdani FRBS model and the measured output maximum power for the total set of data.

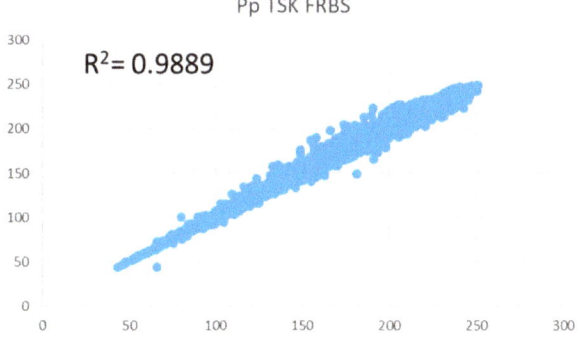

Figure 10. A regression analysis between the predicted output maximum power in the TSK FRBS model and the measured output maximum power for the total set of data.

From the analysis of the experimental results, we note the following observations:

(a) An improvement in the *RMSE* and R^2 of results obtained with TSK FRBS training with ANFIS and with Mamdani FRBS using the Ad Hoc data-driven method compared with the results obtained using ANN.

(b) An improvement in the *RMSE* and R^2 of the results obtained with TSK FRBS training with ANFIS compared with the results obtained using Mamdani FRBS with the Ad Hoc data-driven method.
(c) A strong decrease in the number of rules in the knowledge base obtained for TSK FRBS compared with that obtained for Mamdani FRBS. This simplicity results in an increase in the speed of the inference process.

Nevertheless, the use of a Mamdani FRBS has several advantages, among which we can highlight the fact that each rule is a description of a condition-action statement that has a clear interpretation to a human. In this section, we emphasize the meanings of several rules.

The rule that infers the scene in which the HCPV module generates the least amount of power is:

If DIN is A1, Tair is A17, Ws is A6, PW is A12 and AM is A9, then Power is B1.

By analyzing this linguistic rule, and taking into account the proposed uniform partition of the membership functions, composed by 19 fuzzy sets for each meteorological variables and 495 fuzzy sets in the maximum output power variable, we can say that:

The fuzzy set A1 covers values of the variable DNI very close to its minimum measured value (between 0 and 5% of its range of values). The fuzzy set A17 covers values of the variable Tair close to its maximum measured value (between 87 and 92% of its range of values). The fuzzy set A6 cover values of the variable Ws around 1/3 of its range of measured values (between 29 and 34%). The fuzzy set A12 cover values of the variable PW between 60 and 66% of the range of its measured values. The fuzzy set A9 cover values of the variable AM close to the medium value of its range of measured values (between 45 and 50%). The fuzzy set B1 covers values of the variable Power very close to its minimum value measured (between 0 and 0.2% of its range of values). Therefore we can interpret this rule as follows:

If the value taken by the variable DNI is within the set A1 and if the value taken by the variable Tair is within the set A17 and if the value taken by the variable Ws is within the set A6 and if the value taken by the variable PW is within the set A9 and if the value taken by the variable AM is within the set A9 then the value taken by the variable Power will be within the set B1.

In a simplified way, the conditions that generate the least amount of maximum power are:

A very low DNI, very high Tair, medium-low Ws, medium-high PW and medium AM.

The rule that infers the scene in which the HCPV module generates the largest amount of maximum power is:

If DIN is A16 and Tair is A11 and Ws is A8 and PW is A9 and AM is 1, then Power is B495.

As can see by observing this rule, the maximum amount of maximum power will be supplied under the following conditions:

A very high DNI, medium-high Tair, medium Ws, medium PW, and very low AM.

One of the advantages of Mamdani FRBS is that we can deduce the relationships between the input and output variables. If we consider these rules, we realize that there is a clear connection between the DNI input variable and the amount of electric power supplied by the HCPV module.

The relationships between other variables can be deduced by analyzing other rules in the knowledge base. For example:

If DIN is CB17 and Tair is CB17 and Ws is CB3 and PW is CB8 and AM is CB1, then Power is CB441.

If DIN is CB17 and Tair is CB15 and Ws is CB3 and PW is CB8 and AM is CB1, then Power is CB443.

If DIN is CB17 and Tair is CB14 and Ws is CB3 and PW is CB8 and AM is CB1, then Power is CB454.

If DIN is CB17 and Tair is CB13 and Ws is CB3 and PW is CB8 and AM is CB1, then Power is CB457.

Analyzing these linguistic rules, we realize that under these conditions of the DNI, Ws, PW and AM variables (which are the same in the five rules), the maximum power supplied by this HPCV model will increase if the value of the Tair variable decreases.

Throughout the knowledge base, we can find several sets of rules with the same values for four of the five conditions in their antecedents. Then the values taken by the fuzzy sets for the condition that differs and for the consequent allow for the determination of the relationship between the output maximum power and this input variable.

Although the knowledge base obtained for the HCPV model based on TSK FRBS has only 32 linguistics rules, the analysis of these rules does not allow for the interpretation of the relationships between the input and output variables; thus we cannot deduce human knowledge from the examination of the knowledge base.

In the HCPV modules, the input variables are mutually dependent, for this reason it is difficult to find a proper fuzzy partition of the input variables. To increase the precision of the HCPV model based on Mamdani FRBS, we have been proposed a uniform partition of the space of the input variables and their division into a large number of fuzzy sets. This results in a great increase in the number of linguistic rules.

To decrease the number of rules in the knowledge base without decreasing the accuracy of the model, in the future we can propose a non-uniform partition of the space of input variables, which will decrease the complexity of the input-output mapping and therefore allow for increasing the speed of the inference process and the human interpretability of the linguistic rules in the knowledge bases. This new partition must take into account the density of the examples contained in the input-output data set. This solution must increase the granularity of the fuzzy partition in the areas in which there is a higher density of examples for each one of the input variables.

Another aspect to take in account is the execution time in the process of obtaining the model (the KB). As can see by observing the Table 4, the process of training for Mamdani FRBS model is more expensive in terms of CPU time, than the TSK FRBS and the ANN. However this process is off-line and it is done once.

Table 4. Comparison of the execution time (training, validation and test) for the model TSK FRBS training with ANFIS versus Mamdani FRBS obtained with the Ad Hoc data-driven.

Execution Time (units)	Training (h)	Validation (s)	Test (s)
ANN	1/60	2	2
TSK FRBS training with ANFIS	4	2	2
Mamdani FRBS training with Ad Hoc data-driven	216		2

6. Conclusions

In this paper we have proposed the use of two types of FRBS (Mamdani and Takagi-Sugeno-Kant), with the goal of modelling HCPV systems. With the intention of obtaining the best performances of each fuzzy systems we have proposed a learning method: the ANFIS one, for training a TSK FRBS, in order to obtain its KB, and the Ad Hoc data-driven methodology to obtain the KB of the Mamdani FRBS.

From the analysis of the experimental setup and the description of these methods, it is noticed that the proposed modelling method:

(a) They only need simple outdoor measurements.
(b) They do not need an expensive experimental system.
(c) They are valid for any climatic conditions.
(d) They do not need any specific simulation software.
(e) They can be applied to the modelling of any model of HCPV module.

These characteristics, which they share with the ANN-based method, allow the building of HCPV models without complex requirements. In addition they allow to obtain the complex relationships that link the electrical behaviour of an HCPV module and its inputs variables, with a high level of accuracy.

On the other hand, from the analysis of the experimental results, it is noticed that:

(a) The use of an HCPV model obtained using TSK FRBS trained with ANFIS produced an improvement in the model accuracy compared to the use of the model obtained using ANN.
(b) The use of an HCPV model obtained using Mamdani FRBS with an Ad Hoc data-driven method produced an improvement in the model accuracy compared to the use of the model obtained using ANN.
(c) The use of an HCPV model obtained using TSK FRBS trained with ANFIS produced an improvement in the model accuracy compared to the use of the model obtained using Mamdani FRBS with an Ad Hoc data-driven method.
(d) The use of an HCPV model obtained using Mamdani FRBS with an Ad Hoc data-driven method allowed for the deduction of the relationship between the input and output variables. However, the other HCPV models did not provide information on the modeled system that was interpretable by a human.

For the future work we propose the following lines:

(a) In Mamdani FRBS modelling:

1. To build a non-uniform partition of the membership functions in the input variables, in order to minimize the number of linguistics rules in the KB obtained by applying Ad Hoc data-driven method.
2. To use other learning methods, as genetic algorithms, in order to obtain good KB.

(b) In TSK FRBS trained with ANFIS modelling: to increase the number of fuzzy sets in the partition of the membership functions of the input variables, in order to increase its accuracy.
(c) In FRBS (Mamdani and TSK) modelling: to increase the number of input variables, with new atmospheric variables, in order to obtain HCPV models with higher accuracy.

Author Contributions: M.A.G-M., A.J.Y.D and F.A.C. collected the related work, M.A.G-M. and A.J.Y.D. proposed the methods used to model the HCPV modules. M.A.G-M., A.J.Y.D and F.A.C. conceived and designed the experiments; J.C-B. and J-A.F-P. performed the experiments; J.C-B. and J-A.F-P. analyzed the data; M.A.G-M., J.C-B. and J-A.F-P. wrote the paper.

Funding: This work forms part of the project "Nuevos conceptos basados en tecnología de concentración fotovoltaica: desarrollo de sistemas de muy alta concentración fotovoltaica" (ENE2013-45242-R) supported by the Spanish Economy Ministry and the European Regional Development Fund/Fondo Europeo de Desarrollo Regional (ERDF/FEDER).

Acknowledgments: The authors would like to thank the support given by the IDEA Solar Energy Research Group of the University of Jaén, for the cession of the data captured by the experimental set-up, which have been used for the elaboration of this paper.

Conflicts of Interest: The authors declare no conflict of interest.

References

1. Eduardo, F.; Pérez-Higueras, P.; Garcia Loureiro, A.J.; Vidal, P.G. Outdoor evaluation of concentrator photovoltaic systems modules from different manufacturers: First results and steps. *Prog. Photovolt. Res. Appl.* **2013**, *21*, 693–701. [CrossRef]
2. Siefer, G.; Abbot, P.; Baur, C.; Schleg, T. Determination of the temperature coefficients of various IIIeV solar cells. In Proceedings of the 20th European Photovoltaic Solar Energy Conference, Barcelona, Spain, 6–10 June 2005.
3. Siefer, G.; Baur, C.; Meusel, M.; Dimroth, F.; Bett, A.W.; Warta, W. Influence of the simulator spectrum on the calibration of multi-junction solar cells under concentration. In Proceedings of the Conference Record of the Twenty-Ninth IEEE Photovoltaic Specialists Conference, New Orleans, LA, USA, 19–24 May 2002; pp. 836–839.
4. Fernández, E.F.; Loureiro, A.J.G.; Higueras, P.J.P.; Siefer, G. Monolithic III–V triple-junction solar cells under different temperatures and spectra. In Proceedings of the 8th Spanish Conference on Electron Devices, Palma de Mallorca, Spain, 8–11 February 2011.

5. Siefer, G.; Almonacid, F.; Garcı, A.J.; Ferna, E.F. A two subcell equivalent solar cell model for III–V triple junction solar cells under spectrum and temperature variations. *Sol. Energy* **2013**, *92*, 221–229. [CrossRef]
6. Siefer, G.; Bett, A. Analysis of temperature coefficients for III–V multi-junction concentrator cells. *Prog. Photovolt. Res.* **2014**. [CrossRef]
7. Fernández, E.F.; Siefer, G.; Schachtner, M.; García Loureiro, A.J.; Pérez-Higueras, P. Temperature coefficients of monolithic III–V triple-junction solar cells under different spectra and irradiance levels. *AIP Conf. Proc.* **2012**, *1477*, 189–193.
8. Soria-moya, A.; Cruz, F.A.; Fern, E.F.; Rodrigo, P.; Mallick, T.K.; Pedro, P. Performance Analysis of Models for Calculating the Maximum Power of High Concentrator Photovoltaic Modules. *IEEE J.* **2015**, *5*, 947–955. [CrossRef]
9. Fernández, E.F.; Almonacid, F.; Mallick, T.K.; Pérez-Higueras, P. Analytical modelling of high concentrator photovoltaic modules based on atmospheric parameters. *Int. J. Photoenergy* **2015**, *2015*. [CrossRef]
10. Fernández, E.F.; Soria-Moya, A.; Almonacid, F.; Aguilera, J. Comparative assessment of the spectral impact on the energy yield of high concentrator and conventional photovoltaic technology. *Sol. Energy Mater. Sol. Cells* **2016**, *147*, 185–197. [CrossRef]
11. Fernández, E.F.; Almonacid, F.; Ruiz-Arias, J.A.; Soria-Moya, A. Analysis of the spectral variations on the performance of high concentrator photovoltaic modules operating under different real climate conditions. *Sol. Energy Mater. Sol. Cells* **2014**, *127*, 179–187. [CrossRef]
12. Shanks, K.; Senthilarasu, S.; Mallick, T.K. Optics for concentrating photovoltaics: Trends, limits and opportunities for materials and design. *Renew. Sustain. Energy Rev.* **2016**, *60*, 394–407. [CrossRef]
13. Fernández, E.F.; Almonacid, F.; Rodrigo, P.; Pérez-Higueras, P. Calculation of the cell temperature of a high concentrator photovoltaic (HCPV) module: A study and comparison of different methods. *Sol. Energy Mater. Sol. Cells* **2014**, *121*, 144–151. [CrossRef]
14. Rodrigo, P.; Fernández, E.; Almonacid, F.; Pérez-Higueras, P.J. Review of methods for the calculation of cell temperature in high concentration photovoltaic modules for electrical characterization. *Renew. Sustain. Energy Rev.* **2014**, *38*, 478–488. [CrossRef]
15. Almonacid, F.; Pérez-Higueras, P.J.; Fernández, E.F.; Rodrigo, P. Relation between the cell temperature of a HCPV module and atmospheric parameters. *Sol. Energy Mater. Sol. Cells* **2012**, *105*, 322–327. [CrossRef]
16. Ruiz-Arias, J.A.; Quesada-Ruiz, S.; Fernández, E.F.; Gueymard, C.A. Optimal combination of gridded and ground-observed solar radiation data for regional solar resource assessment. *Sol. Energy* **2015**, *112*, 411–424. [CrossRef]
17. Rodrigo, P.; Fernández, E.; Almonacid, F.; Pérez-Higueras, P.J. Models for the electrical characterization of high concentration photovoltaic cells and modules: A review. *Renew. Sustain. Energy Rev.* **2013**, *26*, 752–760. [CrossRef]
18. ASTM International. *Standard Test Method for Rating Electrical Performance of Concentrator Terrestrial Photovoltaic Modules and Systems Under Natural Sunlight*; E2527-09; ASTM International: West Conshohocken, PA, USA, 2009.
19. Peharz, G.; Ferrer Rodríguez, J.P.; Siefer, G.; Bett, A.W. A method for using CPV modules as temperature sensors and its application to rating procedures. *Sol. Energy Mater. Sol. Cells* **2011**, *95*, 2734–2744. [CrossRef]
20. King, D.L.; Boyson, W.E.; Kratochvil, J.A. *Photovoltaic Array Performance Model*; Paper nr. SAND2004-3844; Sandia National Laboratories: Albuquerque, NM, USA; Livermore, CA, USA, 2004.
21. Whitaker, C.M.; Townsend, T.U.; Newmiller, J.D.; King, D.L.; Boyson, W.E.; Kratochvil, J.A.; Collier, D.E.; Osborn, D.E. Application and validation of a new PV performance characterization method. In Proceedings of the Conference Record of the Twenty Sixth IEEE Photovoltaic Specialists Conference, Anaheim, CA, USA, 29 September–3 October 1997; pp. 1253–1256. [CrossRef]
22. Gerstmaier, T.; Van Riesen, S.; Gombert, A.; Mermoud, A.; Lejeune, T.; Duminil, E. Software modeling of FLATCON© CPV systems. *AIP Conf. Proc.* **2010**, *1277*, 183–186.
23. Nishioka, K.; Takamoto, T.; Agui, T.; Kaneiwa, M. Annual output estimation of concentrator photovoltaic systems using high-efficiency InGaP/InGaAs/Ge triple-junction solar cells based on experimental solar cell's. *Sol. Energy Mater.* **2006**, *90*, 57–67. [CrossRef]
24. Aronova, E.S.; Grilikhes, V.A.; Shvarts, M.Z.; Timoshina, N.H. On correct estimation of hourly power output of solar photovoltaic installations with MJ SCs and sunlight concentrators. In Proceedings of the Conference Record of the IEEE Photovoltaic Specialists Conference, San Diego, CA, USA, 11–16 May 2008.

25. Kinsey, G.S.; Stone, K.; Garboushian, V. Energy prediction of Amonix solar power plants. *Proc. SPIE* **2010**, *7769*, 77690B. [CrossRef]
26. Verlinden, P.J.; Lasich, J.B. Energy rating of Concentrator PV systems using multi-junction III–V solar cells. In Proceedings of the 2008 33rd IEEE Photovoltaic Specialists Conference, San Diego, CA, USA, 11–16 May 2018; pp. 1–6. [CrossRef]
27. Martínez, M.; Antón, I.S. Prediction of PV concentrators energy production: Influence of wind in the cooling mechanisms. First steps. In Proceedings of the 4th International Conference on Solar Concentrators for the Generation of Electricity or Hydrogen, Scottsdale, AR, USA, 1–5 May 2007.
28. Chan, N.L.A.; Young, T.B.; Brindley, H.E.; Ekins-Daukes, N.J.; Araki, K.; Kemmoku, Y.; Yamaguchi, M. Validation of energy prediction method for a concentrator photovoltaic module in Toyohashi Japan. *Prog. Photovolt. Res. Appl.* **2013**, *21*, 1598–1610. [CrossRef]
29. Almonacid, F.; Fernández, E.F.; Rodrigo, P.; Pérez-Higueras, P.J.; Rus-Casas, C. Estimating the maximum power of a High Concentrator Photovoltaic (HCPV) module using an Artificial Neural Network. *Energy* **2013**, *53*, 165–172. [CrossRef]
30. Steiner, M.; Siefer, G.; Hornung, T.; Peharz, G.; Bett, A.W. YieldOpt, a model to predict the power output and energy yield for concentrating photovoltaic modules. *Prog. Photovolt. Res. Appl.* **2015**, *23*, 385–397. [CrossRef]
31. Almonacid, F.; Mellit, A.; Kalogirou, S. Applications of Anns in the Field of the Hcpv Technology. *High Conc. Photovolt.* **2015**, *190*, 333–351.
32. Cordón, Ó.; Herrera, F.; Hoffmann, F.; MAgdalena, L. *Genetic Fuzzy Systems: Evolutionary Tunning and Learning of Fuzzy Knowledge Bases*; World Scientific: Singapore, 2001; Volume 141, ISBN 9810240163.
33. Duckstein, L.; Bardossy, A. *Fuzzy Rule-Based Modeling with Applications to Geophysical, Biological, and Engineering Systems*; CRC Press: Boca Raton, FL, USA, 1995; Volume 8.
34. Pedrycz, W. *Fuzzy Modelling: Paradigms and Practice*; Springer: Berlin/Heidelberg, Germany, 1996; ISBN-13 978-1-4612-8589-2.
35. Sugeno, M.; Yasukawa, T. A Fuzzy-Logic-Based Approach to Qualitative Modeling. *IEEE Trans. Fuzzy Syst.* **1993**, *1*, 7–31. [CrossRef]
36. Takagi, T.; Sugeno, M. Fuzzy Identification of Systems and Its Applications to Modeling and Control. *IEEE Trans. Syst. Man Cybern.* **1985**, *15*, 116–132. [CrossRef]
37. Mamdani, E. Application of fuzzy algorithms for control of simple dynamic plant. *Proc. Inst. Electr. Eng.* **1974**, *121*, 1585. [CrossRef]
38. Mamdani, E.; Assilian, S. An experiment in linguistic synthesis with a fuzzy logic controller. *Int. J. Man Mach. Stud.* **1975**, *7*, 1–13. [CrossRef]
39. Jang, J.S.R. ANFIS: Adaptive-Network-Based Fuzzy Inference System. *IEEE Trans. Syst. Man Cybern.* **1993**, *23*, 665–685. [CrossRef]
40. Cordón, O.; Herrera, F. A proposal for improving the accuracy of linguistic modeling. *IEEE Trans. Fuzzy Syst.* **2000**, *8*, 335–344. [CrossRef]
41. Wang, L.; Mendel, J. Generating fuzzy rules by learning from examples. *IEEE Trans. Syst. Man* **1992**, *22*, 1414–1427. [CrossRef]
42. Casillas, J.; Cordón, O.; Herrera, F. COR: A methodology to improve ad hoc data-driven linguistic rule learning methods by inducing cooperation among rules. *IEEE Trans. Syst. Man Cybern. Part B Cybern.* **2002**, *32*, 526–537. [CrossRef]
43. Pérez-Higueras, P.; Fernández, E.F. *High Concentrator Photovoltaics: Fundamentals, Engineering and Power Plants*; Green Energy and Technology; Springer International Publishing: Berlin/Heidelberg, Germany, 2015; ISBN 9783319150390.
44. Hunt, K.; Sbarbaro, D.; Żbikowski, R.; Gawthrop, P. Neural networks for control systems—A survey. *Automatica* **1992**, *28*, 1083–1112. [CrossRef]
45. Patra, J.C.; Maskell, D.L. Estimation of dual-junction solar cell characteristics using neural networks. In Proceedings of the 2010 35th IEEE Photovoltaic Specialists Conference, Honolulu, HI, USA, 20–25 June 2010; pp. 002709–002713. [CrossRef]
46. Patra, J. Neural network-based model for dual-junction solar cells. *Prog. Photovolt. Res.* **2011**, *19*, 33–44. [CrossRef]
47. Patra, J. Chebyshev neural network-based model for dual-junction solar cells. *IEEE Trans. Energy Convers.* **2011**. [CrossRef]

48. Patra, J.C.; Maskell, D.L. Modeling of multi-junction solar cells for estimation of EQE under influence of charged particles using artificial neural networks. *Renew. Energy* **2012**, *44*, 7–16. [CrossRef]
49. Fernández, E.F.; Almonacid, F.; Garcia-Loureiro, A.J. Multi-junction solar cells electrical characterization by neuronal networks under different irradiance, spectrum and cell temperature. *Energy* **2015**, *90*, 846–856. [CrossRef]
50. Fernández, E.F.; Almonacid, F. Spectrally corrected direct normal irradiance based on artificial neural networks for high concentrator photovoltaic applications. *Energy* **2014**, *74*, 941–949. [CrossRef]
51. Fernández, E.F.; Almonacid, F. A new procedure for estimating the cell temperature of a high concentrator photovoltaic grid connected system based on atmospheric parameters. *Energy Convers. Manag.* **2015**, *103*, 1031–1039. [CrossRef]
52. Rivera, A.J.J.; García-domingo, B.; del Jesus, M.J.; Aguilera, J.; Jesus, M.J.; Aguilera, J.; del Jesus, M.J.; Aguilera, J. Characterization of Concentrating Photovoltaic modules by cooperative competitive Radial Basis Function Networks. *Expert Syst. Appl.* **2013**, *40*, 1599–1608. [CrossRef]
53. Almonacid, F.; Fernández, E.F.; Mallick, T.K.; Pérez-Higueras, P.J. High concentrator photovoltaic module simulation by neuronal networks using spectrally corrected direct normal irradiance and cell temperature. *Energy* **2015**, *84*, 336–343. [CrossRef]
54. García-Domingo, B.; Piliougine, M.; Elizondo, D.; Aguilera, J. CPV module electric characterisation by artificial neural networks. *Renew. Energy* **2015**, *78*, 173–181. [CrossRef]
55. Almonacid, F.; Rus, C.; Hontoria, L.; Fuentes, M.; Nofuentes, G. Characterisation of Si-crystalline PV modules by artificial neural networks. *Renew. Energy* **2009**, *34*, 941–949. [CrossRef]
56. Almonacid, F.; Rus, C.; Pérez, P.; Hontoria, L. Estimation of the energy of a PV generator using artificial neural network. *Renew. Energy* **2009**, *34*, 2743–2750. [CrossRef]
57. Almonacid, F.; Rus, C.; Hontoria, L.; Muñoz, F. Characterisation of PV CIS module by artificial neural networks. A comparative study with other methods. *Renew. Energy* **2010**, *35*, 973–980. [CrossRef]
58. Almonacid, F.; Rus, C.; Pérez-Higueras, P.; Hontoria, L. Calculation of the energy provided by a PV generator. Comparative study: Conventional methods vs. artificial neural networks. *Energy* **2011**, *36*, 375–384. [CrossRef]
59. Almonacid, F.; Rodrigo, P.; Hontoria, L. Generation of ambient temperature hourly time series for some Spanish locations by artificial neural networks. *Renew. Energy* **2013**, *51*, 285–291. [CrossRef]
60. Lee, C.D.; Huang, H.C.; Yeh, H.Y. The development of sun-tracking system using image processing. *Sensors (Switzerland)* **2013**, *13*, 5448–5459. [CrossRef] [PubMed]
61. Alata, M.; Al-Nimr, M.; Qaroush, Y. Developing a multipurpose sun tracking system using fuzzy control. *Energy Convers. Manag.* **2005**, *46*, 1229–1245. [CrossRef]
62. Choi, J.S.; Kim, D.Y.; Park, K.T.; Choi, C.H.; Chung, D.H. Design of fuzzy controller based on PC for solar tracking system. In Proceedings of the ICSMA 2008—International Conference on Smart Manufacturing Application, Gyeonggi-do, Korea, 9–11 April 2008; pp. 508–513.
63. Taherbaneh, M.; Fard, H.G.; Rezaie, A.H.; Karbasian, S. Combination of fuzzy-based maximum power point tracker and sun tracker for deployable solar panels in photovoltaic systems. In Proceedings of the IEEE International Conference on Fuzzy Systems, London, UK, 23–26 July 2007; pp. 1–6.
64. Yousef, H. Design and implementation of a fuzzy logic computer-controlled sun tracking system. In Proceedings of the IEEE International Symposium on Industrial Electronics (Cat. No.99TH8465), Bled, Slovenia, 12–16 July 1999; Volume 3, pp. 1030–1034. [CrossRef]
65. Yeh, H.-Y.; Lee, C.-D. The Logic-Based Supervisor Control for Sun-Tracking System of 1 MW HCPV Demo Plant: Study Case. *Appl. Sci.* **2012**, *2*, 100–113. [CrossRef]
66. Chi, Z.; Yan, H.; Pham, T. *Fuzzy Algorithms: With Applications to Image Processing and Pattern Recognition*; World Scientific: Singapore, 1996.
67. Hirota, K. *Industrial Applications of Fuzzy Technology*; Springer: Berlin/Heidelberg, Germany, 1993.
68. Alcalá, R.; Casillas, J.; Cordón, O.; Herrera, F.; Zwir, S.J.I. Techniques for Learning and Tuning Fuzzy Rule-Based Systems for Linguistic Modeling and their Application. Available online: https://www.researchgate.net/publication/239667434_Techniques_for_Learning_and_Tuning_Fuzzy_Rule-Based_Systems_for_Linguistic_Modeling_and_their_Application_E (accessed on 30 January 2019).
69. Zadeh, A.L. Fuzzy Sets. *Inf. Control* **1965**, *8*, 338–353. [CrossRef]

70. Delgado, M.; Gómez-Skarmeta, A.F.; Martín, F. A fuzzy clustering-based rapid prototyping for fuzzy rule-based modeling. *IEEE Trans. Fuzzy Syst.* **1997**, *5*, 223–233. [CrossRef]
71. Yoshinari, Y.; Pedrycz, W.; Hirota, K. Construction of fuzzy models through clustering techniques. *Fuzzy Sets Syst.* **1993**, *54*, 157–165. [CrossRef]
72. Cordon, O.; Herrera, F. A General Study on Genetic Fuzzy Systems. Available online: https://sci2s.ugr.es/sites/default/files/ficherosPublicaciones/0160_gfs-1995-33-57.pdf (accessed on 30 January 2019).
73. Cordón, O.; Herrera, F.; Lozano, M. A classified review on the combination fuzzy logic-genetic algorithms bibliography: 1989–1995. In *Genetic Algorithms and Fuzzy Logic Systems. Soft Computing Perspectives*; World Scientific: Singapore, 1997; pp. 209–241.
74. Muñoz-Expósito, J.E.; García-Galán, S.; Ruiz-Reyes, N.; Vera-Candeas, P. Adaptive network-based fuzzy inference system vs. other classification algorithms for warped LPC-based speech/music discrimination. *Eng. Appl. Artif. Intell.* **2007**, *20*, 783–793. [CrossRef]
75. Piri, J.; Kisi, O. Modelling solar radiation reached to the Earth using ANFIS, NN-ARX, and empirical models (Case studies: Zahedan and Bojnurd stations). *J. Atmos. Solar-Terrestrial Phys.* **2015**, *123*, 39–47. [CrossRef]
76. Prasad, K.; Gorai, A.K.; Goyal, P. Development of ANFIS models for air quality forecasting and input optimization for reducing the computational cost and time. *Atmos. Environ.* **2016**, *128*, 246–262. [CrossRef]
77. Kharb, R.K.; Shimi, S.; Chatterji, S.; Ansari, M.F. Modeling of solar PV module and maximum power point tracking using ANFIS. *Renew. Sustain. Energy Rev.* **2014**, *33*, 602–612. [CrossRef]
78. Najah, A.; El-Shafie, A.; Karim, O.A.; El-Shafie, A.H. Performance of ANFIS versus MLP-NN dissolved oxygen prediction models in water quality monitoring. *Environ. Sci. Pollut. Res.* **2014**, *21*, 1658–1670. [CrossRef]
79. Driankov, D.; Hellendoorn, H.; Reinfrank, M. *An Introduction to Fuzzy Control*; Springer: Berlin/Heidelberg, Germany, 1993.
80. MODIS Daily Level-3 Data. Available online: http://gdata1.sci.gsfc.nasa.gov/daac-bin/G3/gui.cgi?instance_id=MODIS_DAILY_L3 (accessed on 24 October 2014).

© 2019 by the authors. Licensee MDPI, Basel, Switzerland. This article is an open access article distributed under the terms and conditions of the Creative Commons Attribution (CC BY) license (http://creativecommons.org/licenses/by/4.0/).

Article

Modeling Vehicles to Grid as a Source of Distributed Frequency Regulation in Isolated Grids with Significant RES Penetration

Neofytos Neofytou [1], Konstantinos Blazakis [1], Yiannis Katsigiannis [2,*] and Georgios Stavrakakis [1]

[1] School of Electrical and Computer Engineering, Technical University of Crete, University Campus, GR-73100 Chania, Greece; neofytos_uni@yahoo.com (N.N.); konst.blazakis@gmail.com (K.B.); gstavr@electronics.tuc.gr (G.S.)
[2] Department of Environmental and Natural Resources Engineering, Technological Educational Institute of Crete, Romanou 3, GR-73100 Chania, Greece
* Correspondence: katsigiannis@chania.teicrete.gr; Tel.: +30 282-102-3046

Received: 15 January 2019; Accepted: 19 February 2019; Published: 22 February 2019

Abstract: The rapid development of technology used in electric vehicles, and in particular their penetration in electricity networks, is a major challenge for the area of electric power systems. The utilization of battery capacity of the interconnected vehicles can bring significant benefits to the network via the Vehicle to Grid (V2G) operation. The V2G operation is a process that can provide primary frequency regulation services in the electric network by exploiting the total capacity of a fleet of electric vehicles. In this paper, the impact of the plug-in hybrid electric vehicles (PHEVs) in the primary frequency regulation is studied and the effects PHEVs cause in non-interconnected isolated power systems with significant renewable energy sources (RES) penetration. Also it is taken into consideration the requirements of users for charging their vehicles. The V2G operation can be performed either with fluctuations in charging power of vehicles, or by charging or discharging the battery. So an electric vehicle user can participate in V2G operation either during the loading of the vehicle to the charging station, or by connecting the vehicle in the charging station without any further demands to charge its battery. In this paper, the response of PHEVs with respect to the frequency fluctuations of the network is modeled and simulated. Additionally, by using the PowerWorld Simulator software, simulations of the isolated power system of Cyprus Island, including the current RES penetration are performed in order to demonstrate the effectiveness of V2G operation in its primary frequency regulation.

Keywords: vehicle-to-grid (V2G); isolated power system dynamic simulation; primary frequency control (PFC); scheduled charging; aggregator; battery storage; renewable energy systems modeling

---

1. Introduction

Plug-in hybrid electric vehicles (PHEVs) have the ability to recharge their batteries with electricity from an off-board source (such as the electric utility grid). PHEVs as a new type of load and due to their large-scale integration in the power system will have a significant impact on power system operation and planning, especially in isolated power systems [1,2]. The increased load in power systems due to the charging of electric vehicles should be taken into account in the optimal operation of power systems as well as in investment planning. The major challenge for the area of electric power systems is the utilization of battery capacity of the interconnected vehicles during their charging in charging stations [3]. More precisely, PHEVs can act as small mobile energy storage units and at the same time as controllable loads [4]. An integration of PHEVs in the power grid with appropriate communications and information systems can provide ancillary services to the power grid [5]. Frequency regulation is an important ancillary service that PHEVs can provide because the time duration of this service is short (i.e., a few minutes) and moreover can provide great economic rewards to the owners of vehicles especially in isolated power grids with significant RES penetration [6,7]. With V2G operation PHEVs can participate in electricity balancing markets to provide regulation services [8,9]. The frequency regulation service can be offered by vehicle-to-grid (V2G) operation, which achieves bidirectional power flow between PHEVs and a power grid [10]. When the system frequency goes downwards, PHEVs acting as power producers can prevent further frequency drop. On the other hand, PHEVs could absorb the power from the grid to prevent from a further increase in frequency [11,12]. In this paper, we focus as a case study on the V2G control participating in Primary Frequency Control (PFC) at the isolated power system of Cyprus. PHEVs can participate in PFC because the frequency signal is available at any location of a power system where a PHEV is connected. Except for frequency droop control, V2G control strategies (centralized and decentralized V2G control for PFC) include both maintaining the BSOC (Battery Sate of Charge) and achieving charging demand. When the remaining battery energy is low, the customer has to charge the EV to a higher BSOC level. On the other hand, if the residual BSOC is sufficiently high in the day time, the EV customer is generally willing to maintain the BSOC and recharge the EV in the night considering the low off-peak electricity price [2,13]. In our study, we chose a simple V2G strategy that includes only the droop control and the scheduled charging power of PHEV if the driver wants to charge more the vehicle. The actually charging duration is estimated on the basis of the actual plug-in duration. The optimal driving range of PHEVs to achieve the minimum social cost is examined in Reference [14], where the electric driving range is found to be sensitive to factors, such as the battery pack cost and the gasoline price. These aspects are out of the scope of the present paper, thus they are not taken into consideration here.

The main contribution of this paper is the modeling and simulation of the dynamic operation of the isolated insular power system of Cyprus, which is the largest island in Eastern Mediterranean Sea, with significant RES penetration (mainly from wind turbines), based on real data. These data include mainly load demand, wind turbine production, conventional units technical data, as well as data from system frequency response in real disturbances. In this study it is shown that the frequency droop control that the significant PHEVs integration can offer is able to improve remarkably the primary frequency regulation of the Cypriot power system, which is more sensitive to dynamic variations compared to the large interconnected ones.

This paper is organized as follows. In Section 2, the framework of V2G operation to join in the PFC of a power grid is presented. The V2G control that we chose to achieve frequency regulation, as well as scheduled charging is addressed in Section 3. In Section 4, the PHEVs V2G control block for frequency regulation is presented and discussed. Simulations and discussions that illustrate the performances of the PHEVs V2G control block and V2G control are shown in Section 5. Section 6 discusses the key features of the isolated power system of Cyprus. Section 7 contains a brief description of the adopted wind turbine simulation model. In Section 8 the overall simulation model of the isolated Cyprus Island power system, including the significant RES penetration is presented, and the frequency response is simulated and verified under a real scenario that is related to a conventional power unit loss. Section 9 presents the PHEVs fleet which is integrated in the power system of Cyprus, as well as the simulation results of frequency response with and without V2G grid operation for four cases that are related with a sharp increase or decrease of power system load during a peak load day. Moreover, in this section the BSOC of PHEVs and the exchange power from the power grid side is presented for each case. Finally, the conclusions of this paper are presented in Section 10.

2. System Framework of V2G Operation

Figure 1 illustrates the framework of V2G operation for PHEVs to join in the PFC of a power system. As can be seen, an aggregator is needed for the participation of PHEVs in PFC. The V2G aggregator monitors the fleet of vehicles, which are integrated into the power grid. Actually creates an aggregate profile of a virtual power plant, which depends on the number of vehicles that are connected and capable for V2G, every hour of the day. In addition, the power of the virtual power plant depends on the available power and energy that can provide the fleet of vehicles. The aggregator can receive signals/commands by the transmission system operator and then forward them to the fleet of vehicles, according to the power grid requirements.

In a market environment, aggregator should estimate the regulation capability of the PHEVs fleet and choose an optimal bidding strategy. After market clearing, the aggregator will inform the chosen PHEVs to participate in the PFC [15–17].

PHEVs are connected to the power grid with electric vehicle supply equipment through which bidirectional energy exchange can be achieved. The V2G controller makes decisions based on real-time frequency and BSOC sent from the battery management system (BMS), respectively. It is considered that V2G controller has an embedded frequency detection block which monitors the system frequency in real time. The real-time command is produced and sent from the V2G controller to the charger/discharger block. The charger/discharger is a device that controls the power interchange between the power grid and the PHEV battery to suppress frequency fluctuation. Simultaneously undertakes to achieve charging demand.

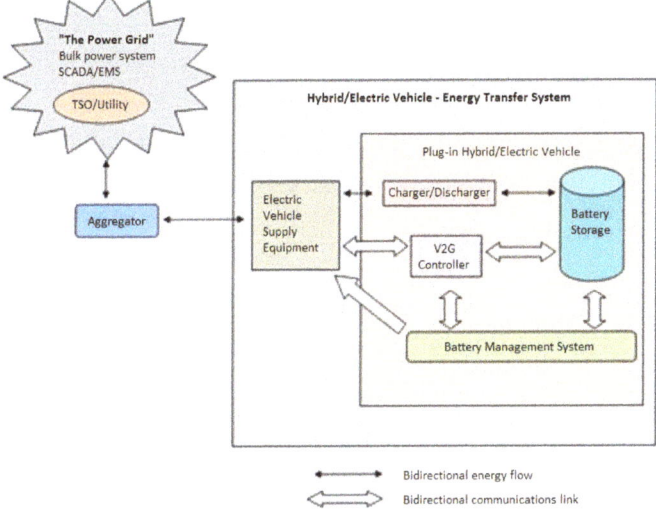

Figure 1. Framework of Vehicle to Grid (V2G) operation participating in primary frequency control (PFC) (TSO: Transmission System Operator, SCADA: Supervisory Control and Data Acquisition, EMS: Energy Management System (EMS)) [18].

The battery management system (BMS) monitors the BSOC and health of the battery. Also, provides an interface between the battery and the rest of the vehicle. With BMS the driver can define some parameters which are associated with the desired SOC, the charging time of the battery and the length of the next trip. Furthermore, the system or the driver of the vehicle must be able to define the maximum discharge depth of the battery [19–21].

3. V2G Control Method for PFC

3.1. Description of V2G Control Method

As illustrated in Figure 1, PHEVs can act as small mobile energy storage units to take part in PFC. In addition, the PHEVs charging demand must be achieved simultaneously. Therefore, frequency regulation and charging demand are two important concerns that need to be handled in the V2G control process. When a PHEV gets to the parking place, the customer will check if the residual BSOC is sufficient for the next trip. If the battery energy at the time of plug-in is sufficient, the PHEV customer usually wants to maintain the residual battery energy. Customers are more willing to charge their vehicles at home, because the electricity price is low off at night. On the other hand if battery energy at the time of plug-in is not sufficient for the next trip, the PHEV customer has to charge the EV to a higher SOC level. Therefore, the requirements from the EV customers can be generally categorized into two types: Maintaining BSOC and achieving charging demand.

In Reference [2], a Decentralized V2G Control (DVC) for PFC is proposed for the two types of PHEVs customers. The proposed DVC method mainly includes Battery SOC Holder (BSH) and Charging with Frequency Regulation (CFR) for PHEVs to participate in PFC. In this paper we use these two types of V2G control, with the exception that we use a constant value that represents the frequency deviation signal. With these assumptions, if the frequency response has the form as in Figure 2 then the constant value that represents the frequency deviation signal is the maximum deviation value, from the nominal frequency. We consider that, the transmission system operator sends commands to the aggregator, according to this frequency error value. In this case, PHEVs must not respond, until the maximum frequency deviation value has reached. Also, if the frequency deviation does not exist or is negligible then PHEVs must not respond.

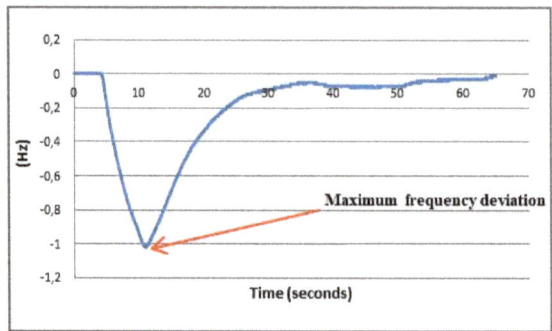

Figure 2. Frequency deviation signal. The maximum deviation value, from the nominal frequency is −1.02 Hz.

3.2. V2G Control to Maintain the Residual Battery Energy

For those PHEVs that need to maintain their SOC levels while joining in PFC, a specific control method is needed to be designed. The BSH which is proposed in Reference [2] is illustrated in Figure 3. We chose the same control method, but with non-adaptive droop and a constant value for frequency error signal. Actually, after plug-out time, the SOC level of PHEV will not be the same as the initial SOC level at the time of plug-in, but very close to it.

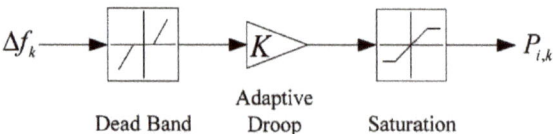

Figure 3. Droop control of the Battery SOC Holder (BSH) for the electric vehicle (EV) charging/discharging power.

The PHEV battery can absorb/inject power from/to a power grid, according to the frequency deviation value and the constant droop K (kW/Hz). For this purpose, a saturation block with upper and lower limits must be included. In addition, a dead-band is added to reduce the charging/discharging operations on the PHEV battery. When the system frequency deviation is out of the predefined dead-band, the power is exchanged between the PHEV and the power grid to suppress frequency fluctuation.

3.3. V2G Control to Achieve Charging Demand

Necessary energy must be supplied to an EV if the residual BSOC is not sufficient for the next trip or the PHEV owner wants to charge the battery overnight. In Reference [2], CFR is proposed to meet charging demand and suppress frequency deviation at the same time. Considering the scheduled charging power, the CFR is presented in Figure 4. As in Section 3.2, we chose the same control method, but with non-adaptive droop and a constant value for frequency error signal. When frequency deviation lies in the predefined dead-band, just scheduled charging power works. Once frequency deviation is out of the dead-band, both scheduled charging power and frequency droop control will work.

The scheduled charging power of a PHEV can be estimated by the following form as a constant:

$$P_i^c = \left(SOC_i^e - SOC_i^{in}\right) \cdot E_i^r / \left(t_i^{out} - t_i^{in}\right), \tag{1}$$

where P_i^c (in kW) is the constant scheduled charging power at the battery side of the ith PHEV for achieving the charging demand, SOC_i^e is the expected state of charge of the ith PHEV battery at plug-out time, SOC_i^{in} is the initial state of charge of the ith EV battery at the time of plug-in, t_i^{out} (in hours) is the plug-out time of the ith PHEV, t_i^{in} is the plug-in time of the ith PHEV, and E_i^r (in kWh) is the rated capacity of the the ith PHEV battery.

The plug-out time, as well as the expected SOC should be provided by the PHEV customer in advance. It should be noted that an EV will not participate in PFC when the scheduled charging power P_i^c is equal to or larger than the maximum charging power.

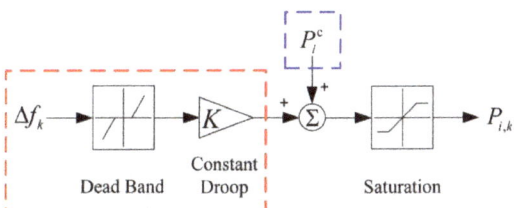

Figure 4. Droop control of the Charging with Frequency Regulation (CFR) (red part represents frequency droop control and the blue part is scheduled charging power).

4. PHEVs with a Simplified V2G Control Block

The simulation model of PHEV is the model proposed in Reference [2], but with a simplified V2G control method presented for the first time here. In PHEV model, real-time BSOC is built to acquire the dynamic change of the battery energy during the V2G operation. As it is shown in Figure 5, the aggregated battery output is controlled based on the input signal Δf_k which represents the frequency deviation at time k. When the control method is applied the input signal converted into power (kW). The estimated power at the connecting point of a charging/discharging device from/to the power grid has the following form:

$$P_{i,k}^p = \begin{cases} \frac{P_{i,k}}{\eta^c}, & (P_{i,k} \geq 0) \\ P_{i,k} \cdot \eta^d, & (P_{i,k} < 0). \end{cases} \tag{2}$$

where η^c is charging efficiency of the PHEVs and where η^d is the discharging efficiency of the PHEVs, $P_{i,k}$ (in kW) is the V2G power at the battery side of the ith EV at time k, and $P_{i,k}^p$ (in kW) is the V2G power at the power grid side of the ith EV at time k.

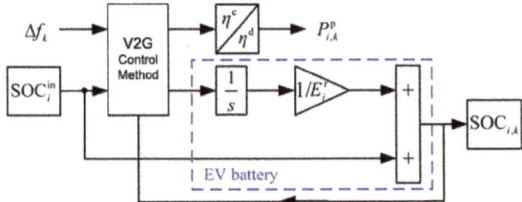

Figure 5. Plug-in hybrid electric vehicles (PHEV) V2G control block for frequency regulation (blue part represents the battery of the vehicle).

The energy variation of the ith PHEV battery during the charging/discharging process can be estimated by the integration of the V2G power (in kW) at the battery side at time period [0–k] as [22–24]:

$$\Delta E_i = \int_0^k P_i(k) dk. \qquad (3)$$

The BSOC usually expressed as a percentage of the rated capacity of the battery, is defined as the available capacity of the battery. The BSOC at time k has the following form:

$$SOC_{i,k} = SOC_i^{in} + \frac{1}{E_i^r} \Delta E_i. \qquad (4)$$

where $SOC_{i,k}$ is the state of charge of the ith PHEV battery at time k, and SOC_i^{in} is the initial state of charge of the ith PHEV battery at the time of plug-in.

5. Test of PHEV Block with V2G Control Application

5.1. Simulation Parameters Values

The PHEV model was implemented by writing code in the Matlab environment. The code is initialized with some parameters which are related to PHEVs and V2G control. The simulation parameters are listed in Table 1. The module of V2G control method was implemented as a function, which performs the operation of the V2G control system. The input signal is the frequency of the power system and is delayed considering the V2G activation and communication delays. The frequency deviation signal is estimated by the code. Afterwards the operation which is described in Section 4 is performed.

Table 1. PHEVs and V2G simulation parameters [2].

Parameter	Measurement Unit	Value
PHEV number		15000
Maximum V2G droop	kW/Hz	3.2
Battery capacity	kWh	32
Maximum SOC	pu	0.1
Minimum SOC	pu	0.9
Maximum V2G power	kW	7
Frequency dead band	Hz	[−0.05, 0.05]
Charging/discharging efficiency	pu	0.92/0.92
LFC delay	s	4

5.2. Simulation Scenarios and Discussion of Results

In the first two simulation scenarios, an input frequency signal with negative frequency deviation is considered. For the other simulation scenarios, an input signal with a positive frequency deviation is considered. The two input signals are illustrated in Figures 6 and 7. In simulation scenarios in which users want to maintain their BSOC level, the simulation time is 80 s in order to study the response of charging power and BSOC of PHEVs, before and after the frequency deviation. In simulation scenarios in which users want to charge their vehicles to a higher BSOC level, the simulation time is three hours (10,800 s).

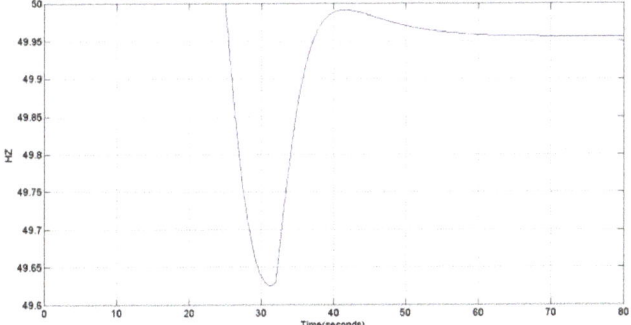

Figure 6. Frequency response signal with negative deviation.

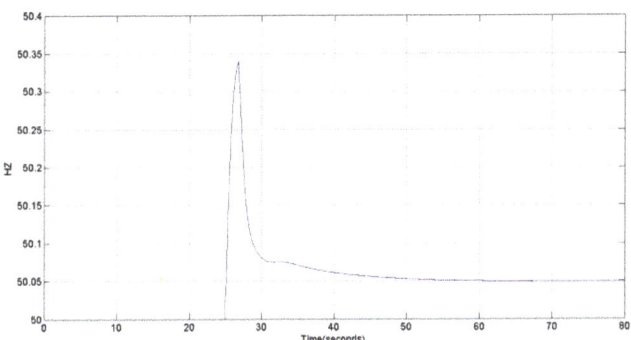

Figure 7. Frequency response signal with positive deviation.

In the first case, PHEVs users need to increase their BSOC level from 50% to 80%. Figure 8 shows the real time SOC and the total V2G power at the power grid side. PHEVs are charging during simulation time. In the 29th second the charging rate is reduced, because a part of charging power is used for power grid needs. Therefore, the scheduled charging power is reduced from 52.17 MW to 32.66 MW. In the 41st second both scheduled charging power and charging rate are reset to their initial values, because the frequency dead band is activated.

In the second case, PHEVs users want to maintain their BSOC level to 80%. As shown in Figure 9, in the 29th second the BSOC of vehicles are reduced, because a part of their stored energy is used for power grid needs. In the 41st second the PHEVs discharging is interrupted, because the frequency dead band is activated. In Figure 9 the initial BSOC of PHEVs is 80% and after the simulation time, the BSOC has reduced by 0.02% of the initial state of charge. Users cannot realize this minimal reduction of BSOC. Therefore, we consider that the state of charge is maintained at its initial level. PHEVs batteries are not discharged significantly because the maximum V2G droop has a small value. If the maximum V2G droop has a large value, the PHEVs batteries will be discharged significantly and the BSOC, after plug-out time, will not approximate the initial BSOC.

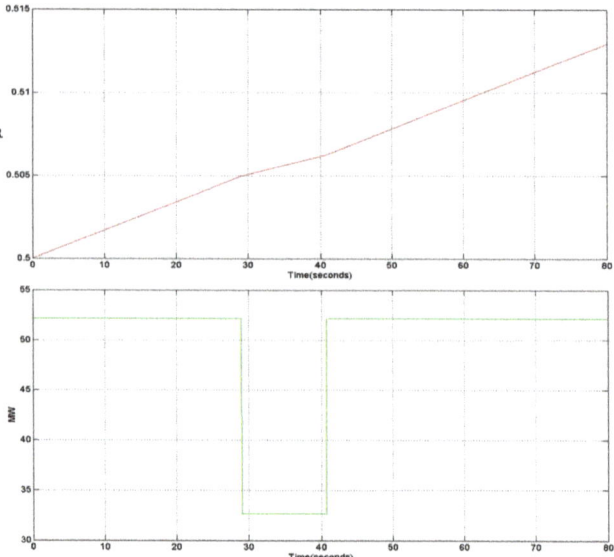

Figure 8. Scenario 1-Red line represents the real-time Battery Sate of Charge (BSOC) and the green line is the total V2G power at the grid side.

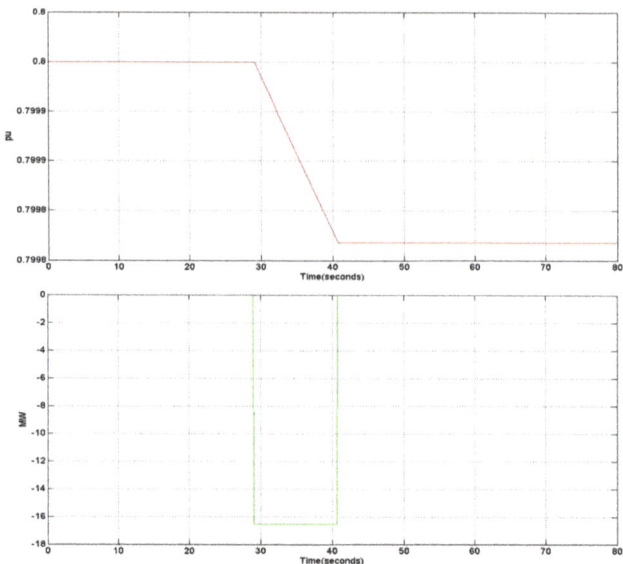

Figure 9. Scenario 2-Red line represents the Real-time BSOC and the green line is the total V2G power at the grid side. PHEVs are discharged in order to reduce the frequency deviation.

In the next two cases, we consider that users have the same desires as the first and second case. For input signal, we set the frequency response that is illustrated in Figure 7. As shown in Figure 10 the charging rate is increased in the 29th second, because the PHEVs absorb more energy from the power grid, in order to reduce the frequency deviation. The scheduled charging power of PHEVs is increased between the 29th and 41st second and after that is reset to its initial value, because the frequency dead band is activated.

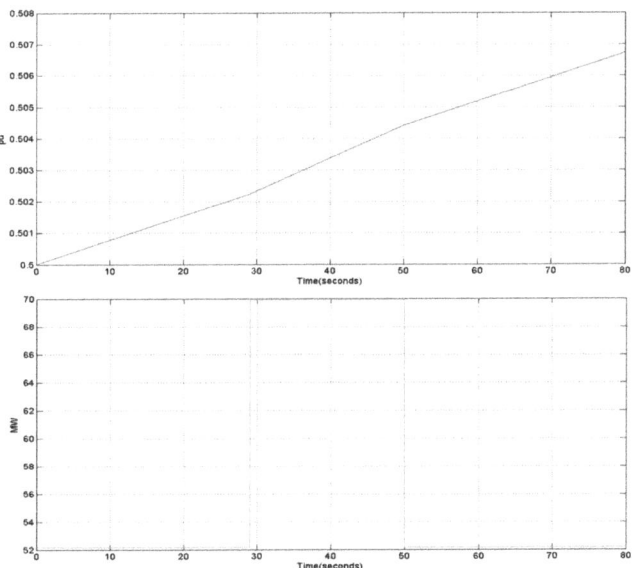

Figure 10. Scenario 3-Red line represents the Real-time BSOC and the green line is the total V2G power at the grid side. PHEVs absorb more energy from the power grid, in order to reduce the frequency deviation.

In Figure 11, the PHEVs charge their batteries in order to reset the frequency to 50 Hz. When the frequency deviation is negligible, the frequency dead band is activated and the PHEVs do not react with the power grid. After the simulation time, the BSOC has increased by 0.02% of the initial state of charge. Therefore, we consider that the state of charge is maintained at its initial level.

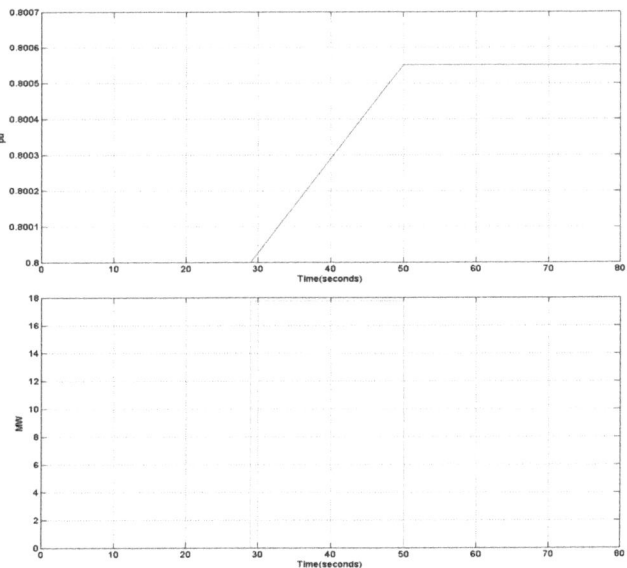

Figure 11. Scenario 4-Red line represents the Real-time BSOC and the green line is the total V2G power at the grid side. PHEVs are charged in order to reduce the frequency deviation.

6. Presentation of the Isolated Power System of Cyprus

Cyprus Island has a small isolated power network running at 50 Hz. The generation system consists of three power stations located at the coast side of the island. The first power station is located in Vasilikos and is composed by 3 × 130 MW Steam Turbine Units (ST), 2 × 220 MW Combined Cycle Gas Turbine Units (CCGT) and one 37.5 MW Gas Turbine (GT) unit. The second power station is located in Dhekelia and consists of 6 × 60 MW ST Units and one Internal Combustion Engine (ICE) of 100 MW. The third power station is located in Moni and consists of 4 × 37.5 MW GT Units. Thus, today the total conventional installed capacity is 1477.5 MW [25]. In addition, renewable energy sources are included in the generation system of Cyprus. Six wind parks with a total installed capacity of 157.5 MW, photovoltaic plants with 77 MW installed capacity and a biomass plant of 10 MW installed capacity, are connected to the Cyprus Island power system [26].

The isolated transmission system of Cyprus is operated at 132 kV and interconnects the major cities and big loads with the three power stations of the island. The interconnections are achieved by using overhead and underground cables. For the year 2017, the maximum demand of the Cyprus Island was recorded to be 1108 MW (during summer) and the minimum demand 310 MW was recorded (during spring) [27].

7. Wind Turbines Simulation Model

For the simulation of wind turbines in Cyprus Island, a Type 2 variable rotor resistance induction generator model was considered. This model consists of the following components (Figure 12): (a) Induction generator, (b) wind turbine, (c) pseudo governor, and (d) rotor resistance controller.

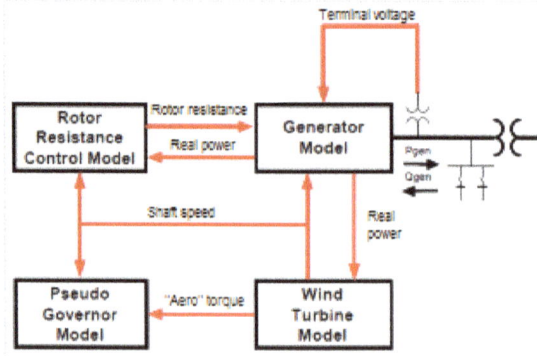

Figure 12. Wind turbine simulation model (Type 2) for Cyprus Island [28].

For the generator, a standard induction generator is used excluding its inertial equation. For the wind turbine, the inertial model of the wind turbine-generator is used, in which the stiffness constant is a function of the first shaft torsional resonant frequency. The pseudo governor mode uses two inputs (rotor speed deviation and generator electrical power), whereas its output is the mechanical power on the rotor blade side. The adjustment of the rotor resistance is implemented via the rotor resistance controller. This controller has as inputs the rotor speed and generator electrical power, while the output is the portion of the available rotor resistance that has to be added to the rotor resistance included in the generator module. For more information about the above mentioned models, the reader is referred to [26]. Thus, the installed asynchronous generator wind turbines in the Cypriot power system cannot contribute to virtual inertia ancillary services, which is possible only with modern variable speed wind turbines interfaced through back-to-back converters, completely decoupling their inertia from the grid [29].

8. Simulation Model of the Cyprus Island Isolated Power System Incorporating RES

8.1. Description of the Developed Simulation Model

The simulation model of the isolated power system of Cyprus Island was designed in the PowerWorld Simulator environment. In Figure 13, the one-line diagram simulation model is presented. As shown in the picture, the power units, the RES units and the load of the power system are included in the model.

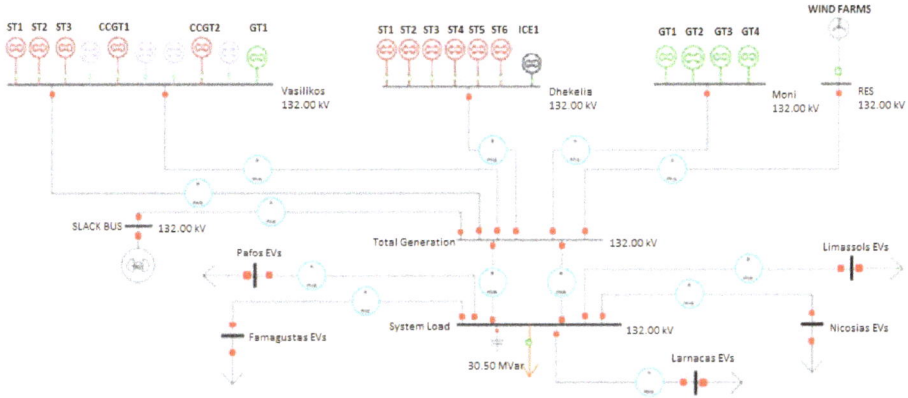

Figure 13. The simulation model of the isolated power system of Cyprus with the load of PHEVs of each city.

The model includes a total of twelve buses. Three buses are used to represent the power stations of Cyprus, and each of them includes the power units of the station. The wind farms of Cyprus Island are simulated by using a bus which includes an equivalent wind generator that represents the total wind power production. The biomass generator of 10 MW is not considered in our analysis because its capacity (10 MW) is negligible compared to the total capacity of conventional generation units (1477.5 MW). Regarding photovoltaics, their effect to frequency stability is also negligible, as their current annual penetration in total electricity production is not significant (less than 3%). It has been shown that the effect of photovoltaics at the frequency of the power system is negligible, even for penetrations up to the level of 20% [30]. This penetration level of photovoltaics could be surpassed marginally only in the case that all photovoltaics operate at their peak power and simultaneously the load demand presents its minimum annual value, which is not possible to happen in the current operational status of the Cyprus Island power system.

The total power generation is transferred through transmission lines with a nominal transfer power of 600 MVA. In addition, two parallel transmission lines transfer the total power generation to a bus which includes the total system load. This load represents the load of the system and the losses of transmission lines. The two parallel transmission lines have a nominal transfer power of 1500 MVA. A capacitor is added in order to maintain the reactive power balance on the bus. The load of PHEVs is not considered at this bus.

As shown in Figure 13, in the simulation model there are even five load buses that represent the PHEVs of each city. Each load represents the aggregate load of the charging stations of the city. In this study the internal operation of the charging stations is considered as "active" loads, which can change depending on the signals that they receive from the aggregator. The load buses are connected via transmission lines with negligible resistance in the bus named "System Load".

In this paper the voltage stability of the power system is not studied. All buses of the model have a nominal voltage of 132 kV which is the nominal voltage of the power transmission system of

Cyprus Island. Also, a slack bus is used to balance the active power and reactive power in the system while performing a power flow study. The slack bus is used in power systems to provide for system losses by emitting or absorbing active and/or reactive power to and from the system. If the system load is less than the generation, then the conventional generators reduce their production. On the other hand, if the system load is bigger than the generation, then the conventional generators increase their production.

For each conventional generator of the simulation model, the GENROU machine model is used, which provides a very good approximation of the dynamic behavior of synchronous generators. In addition, the exciter model IEEET1 of IEEE is added in each synchronous generator. For the STs the TGOV1 governor model is used, that is designed to simulate the operation of a simple governor of an ST. For the ICE, DEGOV1 model is used, whereas tor the GT engines the governor model GAST_GE is used.

8.2. Simulation of the Cyprus Island Power System Dynamic Frequency Response

In order to simulate the frequency response of the isolated power system of Cyprus Island, it is necessary to set the parameters of the generators which affect the frequency response of the system. The most important parameters that are needed to set are the following: The inertia of the generator (in the GENROU model denoted by H), the droop of generator R, the gain K of governor control system and the parameters related to the time delay for the shifting of regulators speed, in the case of change of the frequency of the network. These parameters are called time delays and are given in Reference [31].

In Figure 14 is illustrated the simulated and the real frequency response of the system, when a steam turbine is lost. As is shown, the time period that the frequency drops down from its nominal value is approximated very closely. In addition, the frequency is recovered closely to the nominal value inside the real time bounds. As seen in Figure 14, in both curves frequency is not recovered to the nominal value, but very closely of it. This happens because these curves show the frequency value after the PFC, but not after the secondary frequency regulation. The generators of PowerWorld Simulator are able to simulate only the PFC. In the real network, the units which perform the secondary regulation, undertake to restore the frequency to the nominal value. Table 2 includes the load of each unit and the total load of the system at the time where the unit lost. The load of PHEVs is not considered in the system for this simulation.

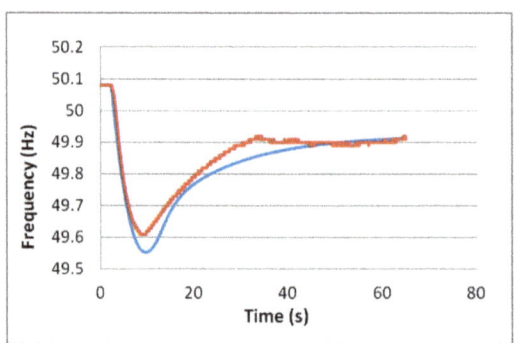

Figure 14. Frequency response of the system when unit is lost. The red curve represents the real frequency response and the blue line represents the simulated frequency response.

Table 2. Composition of units and their loads at the time where the steam unit is lost.

Power Station	Generation Unit	Unit Load (MW)
Vasilikos	Steam turbine-ST1	120
	Steam turbine-ST2	116
	Steam turbine-ST3	0
	Combined cycle-CCGT1	173
	Combined cycle-CCGT2	0
	Gas turbine-GT1	0
Dhekelia	Steam turbine-ST1	0
	Steam turbine-ST2	0
	Steam turbine-ST3	0
	Steam turbine-ST4	30
	Steam turbine-ST5	0
	Steam turbine-ST6	30
	Internal combustion-ICE1	0
Moni	Gas turbine-GT1	0
	Gas turbine-GT2	0
	Gas turbine-GT3	0
	Gas turbine-GT4	0
Wind generation	Wind turbines	37
	Total system load	506

9. Simulations with V2G Operation in the Power Grid

9.1. PHEVs Fleet Estimation

In 2017, the total registered vehicles in Cyprus were 794,464. In order to get an intuitive view about the PHEVs as energy sources, a total number that is equal to 5% of the registered vehicles (39,724 vehicles) is considered. Furthermore, we considered that all PHEVs have a battery with a nominal capacity of 22 kWh. Assuming that all PHEVs are fully charged once a day and that the 80% of their energy potential is available, then PHEVs represent a total energy storage capacity equal to:

$$39{,}724 \text{ PHEVs} \times 22 \text{ kWh} \times 0.8 = 699.14 \text{ MWh}. \tag{5}$$

By using the Matlab and with help of the function "randsample", we distributed the PHEVs of every city in four groups depending on the users SOC requirements which determined randomly. With this way we created a fleet of PHEVs which is presented in Table 3.

Table 3. PHEVs fleet per district.

City/District	Group/PHEVs	SOC_{in} (%)	SOC_{out} (%)	Pc (kW)	Total Pc (MW)
Nicosia	A/7095	37	80	3.153	56.493
	B/4473	24	90	4.84	
	C/2005	33	80	6.446	
	D/1851	49	90	3.006	
Limassol	A/2667	35	80	3.300	37.839
	B/3093	46	90	3.226	
	C/1387	32	90	4.253	
	D/3519	39	90	3.740	
Larnaka	A/2654	13	70	4.180	28.545
	B/1942	17	90	5.353	
	C/1036	29	70	3.006	
	D/840	26	90	4.693	

Table 3. Cont.

City/District	Group/PHEVs	SOC$_{in}$ (%)	SOC$_{out}$ (%)	Pc (kW)	Total Pc (MW)
Famagusta	A/305	38	80	3.080	11.071
	B/1041	12	80	4.986	
	C/762	15	70	4.033	
	D/432	21	80	4.326	
Pafos	A/1294	39	80	3.006	14.633
	B/1941	22	70	3.520	
	C/601	27	70	3.153	
	D/786	45	80	2.566	
Total load on the system (MW)					148.581

9.2. Simulation Results

By using the model of Figure 13 in PowerWorld Simulator along with the V2G control system, a number of simulations were performed, in order to study the V2G operation in PFC. For each simulation scenario the real load data for each unit were used, which were obtained from the TSO of Cyprus Island in order to verify the correct operation of the power system. In addition, we initially add to the system the total load of PHEVs for each one of the five Cyprus towns. Table 3 presents this load, which is divided by the charging efficiency of the vehicles' batteries (148.581 MW/0.92 ≈ 162 MW). Under these circumstances, the system was brought to blackout. This happened because the total production of the units was not enough to undertake the load of vehicles. The result of this action proves that if a large proportion of users try to recharge their vehicles, at a time during the evening, the system will not be able to satisfy the PHEVs demand. To avoid such a serious incident in the power network, forecasting of the PHEVs load should be implemented first. In this study, we should put into operation more conventional generators in order to face this problem. In a real power system, the load distribution in generation units is based on the economic cost of units we used a function from Reference [32], which returns the optimal economical load distribution of the units.

In the following paragraphs, two of the simulations scenarios of this study are presented. In each scenario a table of results is presented, which includes the generation units that are operating and the system load. By joining more units to the power system the operating conditions are changed. For this reason, in the following simulations, we considered that initially the PHEVs are in total a common load on the system, and subsequently we considered that PHEVs are able to perform the V2G operation.

Scenario 1: In this scenario is simulated the frequency response of the system, when the system load is increased sharply. At the time t = 600 seconds, the load at the bus system load is increased by 10%. Initially the load was 1005 MW + 35 MVar and after the fluctuation the load became 1105.5 MW + 35 MVar. The load at the bus System Load was already high before the fluctuation. So we can assume that it represents a moment of a day during summer, where there is a high consumption of electricity. Also, the positive load change may represent a time when most people turn on the air condition systems, as usually happens during the summer noon. Table 4 includes the load of each unit and the total load of the system with and without penetration of PHEVs to the grid. Figure 15 shows the frequency response of the system without V2G operation. As can be seen, the frequency is stabilized around 49.82 Hz after the PFC. In Figure 16, the frequency response, including V2G operation is presented, and the frequency stabilization is improved to 49.9 Hz. As a result, PHEVs are proven to have a positive effect in the PFC of the power grid. In the case of V2G operation, a small frequency vibration due to the voltage vibrations is also observed. The voltage vibrations, are caused by the reaction of the PHEVs with the grid at time t = 611 seconds.

Table 4. Composition of units and there loads with and without PHEVs load in the power system.

Power Station	Generation Unit	Unit Load (MW)	
		W/O PHEVs	With PHEVs
Vasilikos	Steam turbine-ST1	102.4	100
	Steam turbine-ST2	102.4	100
	Steam turbine-ST3	102.4	100
	Combined cycle-CCGT1	165.43	186.5
	Combined cycle-CCGT2	165.43	186.5
	Gas turbine-GT1	4	32
Dhekelia	Steam turbine-ST1	37.799	55
	Steam turbine-ST2	37.799	55
	Steam turbine-ST3	37.799	55
	Steam turbine-ST4	37.799	55
	Steam turbine-ST5	37.799	55
	Steam turbine-ST6	37.799	55
	Internal combustion-ICE1	0	0
Moni	Gas turbine-GT1	20.379	19
	Gas turbine-GT2	20.379	19
	Gas turbine-GT3	20.379	19
	Gas turbine-GT4	0	0
Wind generation	Wind turbines	75	75
Total system load		1005	1167

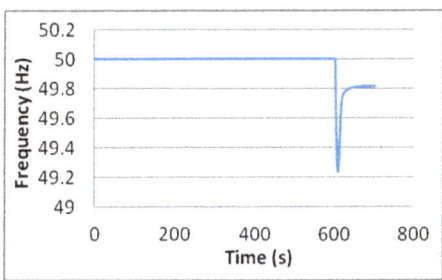

Figure 15. Scenario 1, Frequency response without V2G operation.

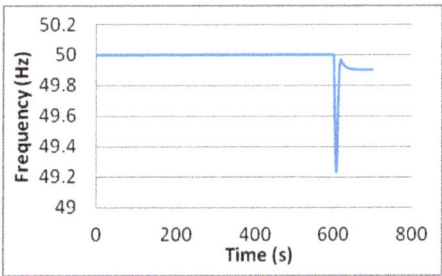

Figure 16. Scenario 1, Frequency response with V2G operation.

In Figure 17, the BSOC of Nicosia's PHEVs is illustrated. The PHEVs batteries are charged all the time of simulation. At time t = 611 seconds the charging rate is decreased in order to support the PFC of the power grid.

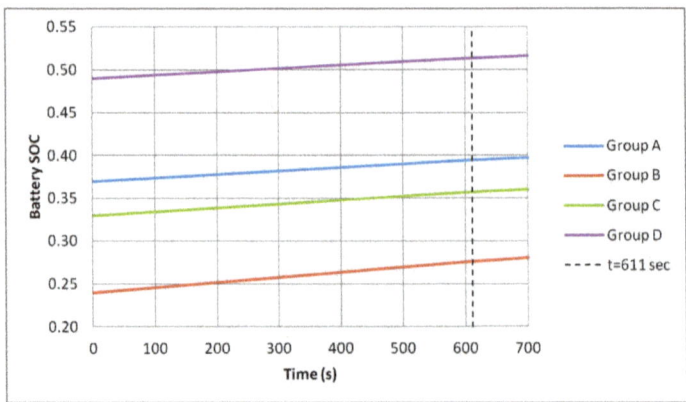

Figure 17. BSOC of Nicosia's PHEVs.

Figure 18 shows the charging power of Nicosia's PHEVs from the grid side. The charging power of each group of vehicles is reduced at time t = 611 seconds and is maintained at a lower level than the initial power, during the PFC of the power grid. When the frequency of the power grid is returned to its nominal value, the charge rate and the charging power of PHEVs will return to their initial values. However this case cannot be studied in this paper because PowerWorld Simulator does not simulate the Secondary Frequency Regulation.

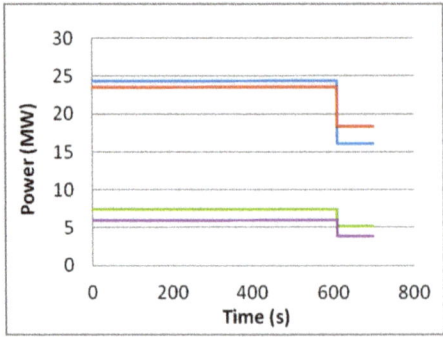

Figure 18. Scenario 1, Charging power of Nicosia's PHEVs. Blue line represents group A, red line for group B, green for group C and purple line for group D.

The considered power grid contains not only PHEVs that their owners desire to charge their batteries, but also PHEVs that their owners desire to maintain their BSOC and to participate in V2G operation simultaneously. In order to study this case, the PHEVs fleet that presented in Table 3 was increased by the addition of 25,000 PHEVs that they are connected into the power grid, and they simultaneously desire to maintain their initial BSOC. Table 5 shows the fleet distribution of these 25,000 PHEVs.

The total load of the system is not changed with the addition of 25,000 PHEVs because the drivers of vehicles, don't desire to charge their cars. So, the 25,000 PHEVs can be considered as backup energy storage units that are connected to power grid. The scenario was implemented with the same load at the System Load bus and at the time t = 600 seconds increased by 10%. The frequency response with V2G operation from the new fleet of PHEVs is presented in Figure 19. As we can see, the frequency is stabilized around 49.95 Hz after the PFC, which is very close to its nominal value. This has happened because the V2G operation was strengthened by the 25,000 PHEVs, which acted as emergency power sources at the time of fluctuation. In this way, the work of the secondary frequency control becomes even easier. A small frequency vibration is observed due to the voltage vibrations.

Table 5. PHEVs fleet of vehicles that desire to maintain their initial BSOC.

City/District	Group/PHEVs	SOC$_{in}$ (%)
Nicosia	A/2000	86
	B/1898	65
	C/1745	74
	D/1857	52
Limassol	A/1236	71
	B/1248	53
	C/1265	89
	D/1251	74
Larnaka	A/1107	59
	B/1157	82
	C/1124	53
	D/1112	57
Famagusta	A/944	69
	B/990	64
	C/1039	58
	D/1027	66
Pafos	A/1016	64
	B/1001	58
	C/986	65
	D/997	61
Total PHEVs	25,000	

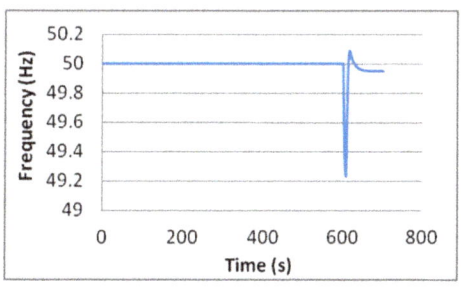

Figure 19. Scenario 1, frequency response with V2G operation, with the additional fleet of PHEVs.

In Figure 20, are illustrated the BSOC and the transmitted power to the grid from Nicosia's PHEVs, which desire to maintained their BSOC to initial level. PHEVs are discharged at the time of fluctuation in order to participate in PFC of the power grid. The discharging rate is the same for all PHEVs because the gain K (kW/Hz) remains the same for all vehicle groups. The BSOC of each vehicle was reduced approximately by 0.13%, which is negligible. This means that the driver of PHEV cannot perceive this minimum discharge of the battery.

At time t = 611 seconds, PHEV batteries are discharged in order to support the power grid. PHEVs discharge their batteries until the frequency stabilized in a constant value. The amount of power which is provided to the power grid from each group of PHEVs depends on the number of vehicles. In addition, the discharge capacity of each vehicle is considered the same for all vehicles, since all vehicles act based on the V2G droop K (kW/Hz) which is the same for all vehicles.

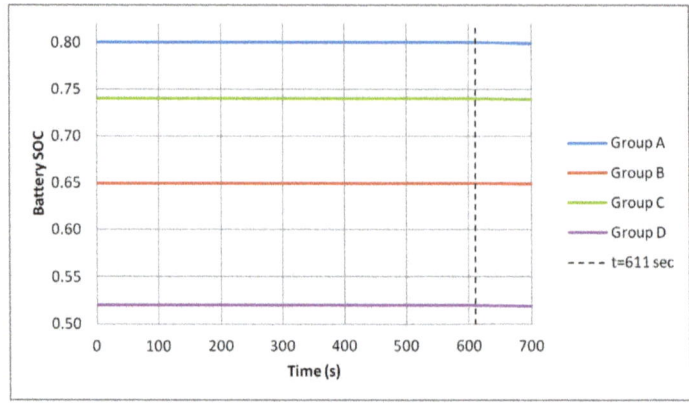

Figure 20. Scenario 1, the BSOC of an additional fleet of Nicosia's PHEVs.

Scenario 2: In this scenario is simulated the frequency response of the system, when the system load is decreased sharply. At the time t = 600 seconds, the load at the System Load bus is decreased by 10%. Initially the load was 1005 MW + 35 MVar and after the fluctuation the load became 904.5 MW + 35 MVar. The total load of the system with and without penetration of PHEVs to the grid is the same as scenario 1. Figure 21 shows the frequency response of the system without V2G operation. As can be seen, the frequency is stabilized around 50.17 Hz after the PFC. In Figure 22, the frequency response, including V2G operation is presented, and the frequency stabilization is improved to 50.11 Hz, which proves again the positive effect in of PHEVs in PFC of the network. In the case of V2G operation, we observe a small frequency vibration due to the voltage vibrations. The voltage vibrations are caused by the reaction of the PHEVs with the grid at time t = 611 seconds.

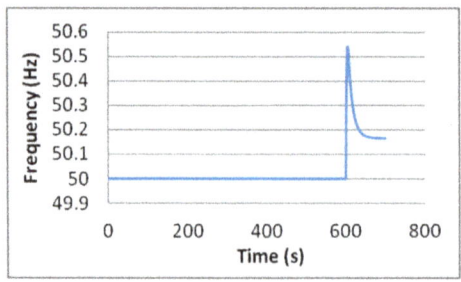

Figure 21. Scenario 2, Frequency response without V2G operation.

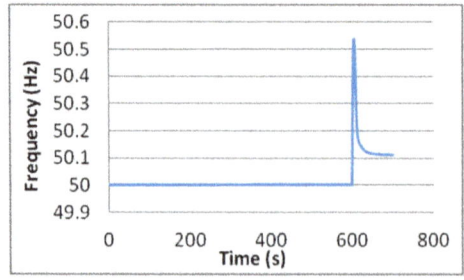

Figure 22. Scenario 2, Frequency response with V2G operation.

In Figure 23 is illustrated the BSOC of Limassol's PHEVs. The PHEVs batteries are charged all the time of simulation. At time t = 611 seconds the charging rate is increased in order to support the PFC of the power grid.

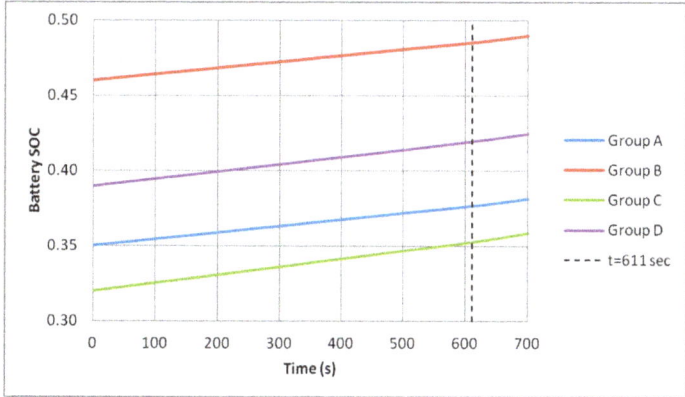

Figure 23. Scenario 2, BSOC of Limassol's PHEVs.

Figure 24 shows the charging power of Limassol's PHEVs from the grid side. The charging power of each group of vehicles is increased at time t = 611 seconds and is maintained at a higher level than the initial power, during the PFC of the power grid. When the frequency of the power grid is returned to its nominal value, then the charge rate and the charging power of PHEVs will return to their initial values.

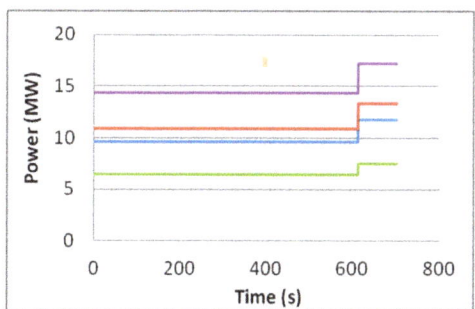

Figure 24. Scenario 2, Charging power of an additional fleet of Limassol's PHEVs. Blue line represents the group A, red line for group B, green for group C and purple line for group D.

As mentioned in scenario 1, in order to study this case, we increased the PHEVs fleet that by addition of 25,000 PHEVs that they are connected into the power grid and desire to maintain their initial BSOC. The frequency response with V2G operation from the new fleet of PHEVs is presented in Figure 25. As can be seen, the frequency is stabilized around 50.07 Hz after the PFC, which is very close to its nominal value. This has happened because the V2G operation was strengthened by the 25,000 PHEVs addition, which acted as emergency power sources at the time of fluctuation. In this way, the operation of the secondary frequency control becomes even easier. A small frequency vibration is observed due to the voltage vibrations.

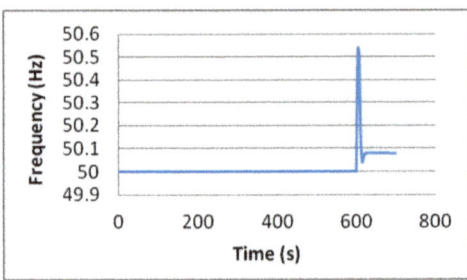

Figure 25. Scenario 2, Frequency response with V2G operation, with the additional fleet of PHEVs.

In Figure 26, the BSOCs from Limassol's PHEVs are illustrated. PHEVs are charged at the time of fluctuation in order to participate in PFC of the power grid. The charging rate is same for all PHEVs because the gain K (kW/Hz) remains the same for all vehicle groups. The BSOC of each vehicle was increased approximately by 0.082% of the nominal capacity of its battery, so the change in BSOC of vehicles is negligible. This means that the driver of PHEV cannot perceive this minimum charge of the battery. At time t = 611 seconds, PHEV batteries are charged in order to support the power grid. PHEVs charge their batteries until the frequency stabilized in a constant value. All vehicles act based on the V2G droop K (kW/Hz) which is considered the same for all vehicles. Figure 27 shows the charging power for the additional fleet in Limassol from the grid side for the four PHEVs groups.

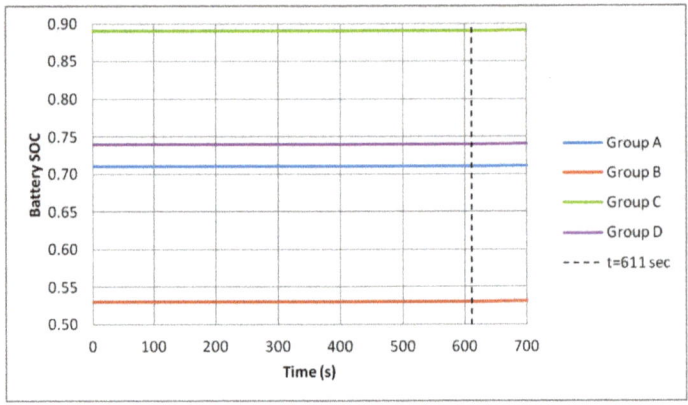

Figure 26. Scenario 2, BSOC of an additional fleet of Limassol's PHEVs.

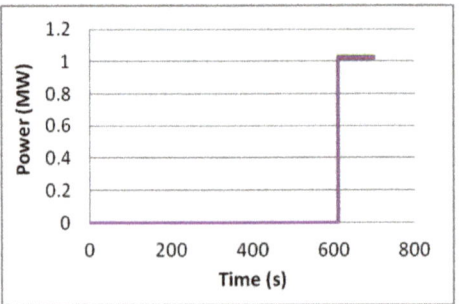

Figure 27. Scenario 2, Charging power of an additional fleet of Limassol's PHEVs. Blue line represents the group A, red line for group B, green for group C and purple line for group D.

It has to be noticed that in the above analysis was considered that all plug-in electric vehicles that were used are PHEVs. However, the category of plug-in electric vehicles includes also battery electric vehicles (BEVs). The diversity of travel demands of different drivers in different days and its influence on designing a proper battery capacity for BEVs are examined in Reference [33]. In the present paper, the impact of the PHEVs fleet as an ancillary system is considered globally and only during the periods that PHEVs are connected to the electricity grid (see Section 9), thus a typical battery capacity is considered for each vehicle, given that is impossible to take into account in a global analysis as the above, proper battery design for each individual vehicle based on the diversity of daily vehicle miles. Additionally, it has to be emphasized that in our study BEVs were not considered as a realistic alternative in the prompt and distant future, because the necessary infrastructure is not envisaged to be available in the Cyprus Island promptly.

10. Conclusions

In this paper, we focused on vehicle to grid (V2G) operation, as well as in the significant benefits by the use of plug-in hybrid electric vehicles (PHEVs) as distributed frequency regulation sources in isolated power systems with significant RES penetration. With the use of appropriate V2G control, the frequency deviation of the system can be suppressed while charging demand is achieved simultaneously. This can be achieved by using the capacity and the stored energy from the batteries of PHEVs. A simple PHEV model was developed in Matlab in order to study this operation and its effects. Moreover, a number of simulations of the isolated power system of Cyprus Island were implemented in PowerWorld Simulator, in order to reveal this operation in a real isolated power system with RES penetration. Through extensive simulations, we observed that after a fluctuation in the power system operation, the frequency dynamic response is closer to the nominal one due to the beneficial V2G operation. Therefore, PHEVs can act as frequency regulation sources in a power system, and especially in an isolated one. A user/owner of PHEV can participate in V2G operation when charging the PHEV in a charging station or simply by the plug-in connection of the vehicle to the network. This may improve the network performance without requiring the installation of new conventional generating units or other costly forms of conventional electric energy spinning reserve. Moreover, the user/owner of PHEV could be paid for these services provided to the power network. Finally, the V2G operation in an isolated power grid incorporating RES, in order to be sustainable, it requires the existence of an appropriate minimum number of electric or hybrid electric vehicles and distributed charging stations. This paper is limited to consider the results achieved from the primary frequency regulation of the Cypriot isolated power system when a significant PHEVs integration occurs in the system, while the secondary frequency regulation procedure is not considered. Nevertheless, this fact does not affect the validity and the importance of the presented results because the most important period for the dynamic stability of the simulated Cypriot isolated power system is the primary frequency regulation duration.

Author Contributions: Conceptualization, G.S. and Y.K.; Methodology, G.S. and Y.K.; Software, N.N. and K.B.; Validation, N.N. and K.B.; Formal Analysis, G.S. and Y.K.; Investigation, G.S. and Y.K.; Resources, G.S. and Y.K.; Data Curation, G.S. and Y.K.; Writing-Original Draft Preparation, N.N., K.B., Y.K. and G.S.; Writing-Review & Editing, G.S. and Y.K.; Visualization, G.S. and Y.K.; Supervision, G.S. and Y.K.; Project Administration, none; Funding Acquisition, none.

Funding: This research received no external funding.

Conflicts of Interest: The authors declare no conflict of interest.

References

1. Jia, H.; Li, X.; Mu, Y.; Xu, C.; Jiang, Y.; Yu, X. Coordinated control for EV aggregators and power plants in frequency regulation considering time-varying delays. *Appl. Energy* **2018**, *210*, 1363–1376. [CrossRef]
2. Liu, H.; Hu, Z.; Song, Y.; Lin, J. Decentralized vehicle-to-grid control for primary frequency regulation considering charging demands. *IEEE Trans. Power Syst.* **2013**, *28*, 3480–3489. [CrossRef]

3. Liu, H.; Hu, Z.; Song, Y.; Wang, J.; Xie, X. Vehicle-to-grid control for supplementary frequency regulation considering charging demands. *IEEE Trans. Power Syst.* **2015**, *30*, 3110–3119. [CrossRef]
4. Yang, J.; Zeng, Z.; Tang, Y.; Yan, J.; He, H.; Wu, Y. Load frequency control in isolated micro-grids with electrical vehicles based on multivariable generalized predictive theory. *Energies* **2015**, *8*, 2145–2164. [CrossRef]
5. Ko, K.S.; Han, S.; Sung, D.K. Performance-based settlement of frequency regulation for electric vehicle aggregators. *IEEE Trans. Smart Grid* **2018**, *9*, 866–875. [CrossRef]
6. Hu, Z.; Zhan, K.; Zhang, H.; Song, Y. Pricing mechanisms design for guiding electric vehicle charging to fill load valley. *Appl. Energy* **2016**, *178*, 155–163. [CrossRef]
7. Zhang, H.; Hu, Z.; Xu, Z.; Song, Y. Optimal planning of PEV charging station with single output multiple cables charging spots. *IEEE Trans. Smart Grid* **2017**, *8*, 2119–2128. [CrossRef]
8. Zhong, J.; He, L.; Li, C.; Cao, Y.; Wang, J.; Fang, B.; Xiao, G. Coordinated control for large-scale EV charging facilities and energy storage devices participating in frequency regulation. *Appl. Energy* **2014**, *123*, 253–262. [CrossRef]
9. Dutta, A.; Debbarma, S. Frequency regulation in deregulated market using vehicle-to-grid services in residential distribution network. *IEEE Syst. J.* **2018**, *12*, 2812–2820. [CrossRef]
10. Krueger, H.; Cruden, A. Modular strategy for aggregator control and data exchange in large scale Vehicle to Grid (V2G) applications. *Energy Procedia* **2018**, *151*, 7–11. [CrossRef]
11. Zhu, X.; Xia, M.; Chiang, H.D. Coordinated sectional droop charging control for EV aggregator enhancing frequency stability of microgrid with high penetration of renewable energy sources. *Appl. Energy* **2018**, *210*, 936–943. [CrossRef]
12. Han, H.; Huang, D.; Liu, D.; Li, Q. Autonomous frequency regulation control of V2G (Vehicle-to-grid) system. In Proceedings of the IEEE Control and Decision Conference (CCDC), Chongqing, China, 28–30 May 2017; pp. 5826–5829.
13. Han, S.; Han, S. Economic feasibility of V2G frequency regulation in consideration of battery wear. *Energies* **2013**, *6*, 748–765. [CrossRef]
14. Kontou, E.; Yin, Y.; Lin, Z. Socially optimal electric driving range of plug-in hybrid electric vehicles. *Transp. Res. Part D Transp. Environ.* **2015**, *39*, 114–125. [CrossRef]
15. Martinenas, S.; Marinelli, M.; Andersen, P.B.; Træholt, C. Implementation and demonstration of grid frequency support by V2G enabled electric vehicle. In Proceedings of the IEEE Power Engineering Conference (UPEC), Cluj-Napoca, Romania, 2–5 September 2014; pp. 1–6.
16. Chukwu, U.C.; Mahajan, S.M. Modeling of V2G net energy injection into the grid. In Proceedings of the IEEE 6th International Conference Clean Electrical Power (ICCEP), Santa Margherita Ligure, Italy, 27–29 June 2017; pp. 437–440.
17. Child, M.; Nordling, A.; Breyer, C. The impacts of high V2G participation in a 100% renewable Aland energy system. *Energies* **2018**, *11*, 2206. [CrossRef]
18. An, K.; Song, K.B.; Hur, K. Incorporating charging/discharging strategy of electric vehicles into security-constrained optimal power flow to support high renewable penetration. *Energies* **2017**, *10*, 729. [CrossRef]
19. Meng, J.; Mu, Y.; Jia, H.; Wu, J.; Yu, X.; Qu, B. Dynamic frequency response from electric vehicles considering travelling behavior in the Great Britain power system. *Appl. Energy* **2016**, *162*, 966–979. [CrossRef]
20. Ota, Y.; Taniguchi, H.; Baba, J.; Yokoyama, A. Implementation of autonomous distributed V2G to electric vehicle and DC charging system. *Electr. Power Syst. Res.* **2015**, *120*, 177–183. [CrossRef]
21. Vachirasricirikul, S.; Ngamroo, I. Robust LFC in a smart grid with wind power penetration by coordinated V2G control and frequency controller. *IEEE Trans. Smart Grid* **2014**, *5*, 371–380. [CrossRef]
22. Masuta, T.; Yokoyama, A. Supplementary load frequency control by use of a number of both electric vehicles and heat pump water heaters. *IEEE Trans. Smart Grid* **2012**, *3*, 1253–1262. [CrossRef]
23. Shimizu, K.; Masuta, T.; Ota, Y.; Yokoyama, A. A new load frequency control method in power system using vehicle-to-grid system considering users' convenience. In Proceedings of the 17th Power System Computation Conference, Stockholm, Sweden, 22–26 August 2011; pp. 1–8.
24. Shimizu, K.; Masuta, T.; Ota, Y.; Yokoyama, A. Load frequency control in power system using vehicle-to-grid system considering the customer convenience of electric vehicles. In Proceedings of the International Conference of Power System Technology, Hangzhou, China, 24–28 October 2010; pp. 1–8.

25. Transmission System Operator Cyprus: Electrical Energy Generation. December 2018. Available online: https://www.dsm.org.cy/en/cyprus-electrical-system/electrical-energy-generation (accessed on 13 January 2019).
26. Cyprus Energy Regulatory Authority (CERA): Production of Electricity Using RES Statistics. Available online: https://www.cera.org.cy/en-gb/ilektrismos/details/statistika-ape (accessed on 13 January 2019).
27. Nikolaidis, P.; Chatzis, S.; Poullikkas, A. Renewable energy integration through optimal unit commitment and electricity storage in weak power networks. *Int. J. Sustain. Energy* **2018**, 1–17. [CrossRef]
28. Ellis, A.; Kazachkov, Y.; Muljadi, E.; Pourbeik, P.; Sanchez-Gasca, J.J. Description and technical specifications for generic WTG models—A status report. In Proceedings of the IEEE Power Systems Conference and Exposition (PSCE), Phoenix, AZ, USA, 20–23 March 2011; pp. 1–8.
29. Tamrakar, U.; Shrestha, D.; Maharjan, M.; Bhattarai, B.; Hansen, T.; Tonkoski, R. Virtual inertia: Current trends and future directions. *Appl. Sci.* **2017**, *7*, 654. [CrossRef]
30. Rahmann, C.; Castillo, A. Fast frequency response capability of photovoltaic power plants: The necessity of new grid requirements and definitions. *Energies* **2014**, *7*, 6306–6322. [CrossRef]
31. Stavrinos, S.; Petoussis, A.G.; Theophanous, A.L.; Pillutla, S.; Prabhakara, F.S. Development of a validated dynamic model of Cyprus Transmission system. In Proceedings of the 7th Mediterranean Conference and Exhibition on Power Generation, Transmission, Distribution and Energy Conversion, Agia Napa, Cyprus, 7–10 November 2010; pp. 1–11.
32. Katsigiannis, Y.A.; Karapidakis, E.S. *The Effect of Pumped Hydro Storage Units Installation on the Operation of the Autonomous Cretan Power System*; Power Systems, Energy Markets and Renewable Energy Sources in South-Eastern Europe; Trivent Publishing: Budapest, Hungary, 2016; pp. 297–308.
33. Li, Z.; Jiang, S.; Dong, J.; Wang, S.; Ming, Z.; Li, L. Battery capacity design for electric vehicles considering the diversity of daily vehicles miles traveled. *Transp. Res. Part C Emerg. Technol.* **2016**, *72*, 272–282. [CrossRef]

© 2019 by the authors. Licensee MDPI, Basel, Switzerland. This article is an open access article distributed under the terms and conditions of the Creative Commons Attribution (CC BY) license (http://creativecommons.org/licenses/by/4.0/).

MDPI
St. Alban-Anlage 66
4052 Basel
Switzerland
Tel. +41 61 683 77 34
Fax +41 61 302 89 18
www.mdpi.com

Energies Editorial Office
E-mail: energies@mdpi.com
www.mdpi.com/journal/energies

www.ingramcontent.com/pod-product-compliance
Lightning Source LLC
LaVergne TN
LVHW071951080526
838202LV00064B/6717